# SPATIAL
# STATISTICS

GeoSpatial Information Modeling
and Thematic Mapping

# SPATIAL STATISTICS

## GeoSpatial Information Modeling and Thematic Mapping

Mohammed A. Kalkhan

CRC Press
Taylor & Francis Group
Boca Raton  London  New York

CRC Press is an imprint of the
Taylor & Francis Group, an **informa** business

CRC Press
Taylor & Francis Group
6000 Broken Sound Parkway NW, Suite 300
Boca Raton, FL 33487-2742

First issued in paperback 2019

ISBN-13: 978-1-4200-6976-1 (hbk)
ISBN-13: 978-0-367-86562-7 (pbk)

**Visit the Taylor & Francis Web site at**
**http://www.taylorandfrancis.com**

**and the CRC Press Web site at**
**http://www.crcpress.com**

# Contents

Preface.........................................................................................................xi
About the Author ...................................................................................... xiii

**1. Geospatial Information Technology** ....................................................... 1
Remotely Sensed Data................................................................................ 1
   Instantaneous Field of View (IFOV) at Nadir
   (Resolution on the Ground) ................................................................. 3
   IKONOS................................................................................................. 4
      Sensor Characteristics ................................................................... 5
      IKONOS Specifications................................................................... 5
   ORBIMAGE (GeoEye) ........................................................................ 5
      OrbView-2 Specifications.............................................................. 6
      OrbView-3 Specifications .............................................................. 6
   QuickBird .............................................................................................. 6
      QuickBird Satellite Sensor Characteristics ................................. 7
   The SPOT (System Probatori d'Observation de la Terre)............... 7
      SPOT-5 Satellite Sensor Characteristics ..................................... 8
   MODIS (Moderate Resolution Imaging Spectroradiometer) ........... 9
      MODIS Overview ......................................................................... 10
      Technical Specifications of MODIS............................................. 10
      MODIS Land Products ................................................................ 10
   ASTER (Advanced Spaceborne Thermal Emission
   and Reflection radiometer) ................................................................ 13
      ASTER Uniqueness ..................................................................... 13
      History of ASTER......................................................................... 14
      Organizational Framework of ASTER ...................................... 14
Active Remotely Sensed Data ................................................................ 15
   Radar.................................................................................................... 15
   Lidar..................................................................................................... 17
      Lidar System Differences ............................................................ 18
      How Does Lidar Work? ............................................................... 18
Derived Remotely Sensed Data ............................................................. 19
   Vegetation Indices.............................................................................. 19
   The Tasseled Cap Transformation................................................... 22
Geographic Information Systems (GIS) ................................................ 25
   Thematic Data Layers........................................................................ 26
Geospatial Data Conversion................................................................... 27
   Using ERDAS-IMAGINE Software ................................................. 27
   Using ARCINFO Software................................................................. 29

Select Area of Interest (Study Site)........................................................31
Topographic Data....................................................................................31
Global Positioning System (GPS)................................................................32
GPS Services ..........................................................................................33
The GPS Satellite System and Facts.......................................................33
GPS Applications ...................................................................................34
References ..................................................................................................35

2. **Data Sampling Methods and Applications**..............................39
Data Representation ...................................................................................39
Data Collection and Source of Errors ........................................................39
Data Types................................................................................................39
Sampling Methods and Applications .........................................................40
Sampling Designs.......................................................................................41
Simple Random Sampling .....................................................................41
Stratified Random Sampling...................................................................42
Systematic Sampling...............................................................................42
Nonaligned Systematic Sample .............................................................44
Cluster Sampling.....................................................................................44
Multiphase (Double) Sampling ..............................................................44
Double Sampling and Mapping Accuracy.................................................45
Pixel Nested Plot (PNP): Case Study.....................................................46
Plot Design..................................................................................................49
Issues........................................................................................................49
Characteristics of Different Plot Shapes ...............................................49
Plot Size ...................................................................................................51
What to Record ...................................................................................51
Issues....................................................................................................51
References ...................................................................................................52

3. **Spatial Pattern and Correlation Statistics** ..............................57
Scale ............................................................................................................57
Spatial Sampling ........................................................................................58
Errors in Spatial Analysis......................................................................58
Spatial Variability and Method of Prediction.......................................58
Spatial Pattern ...........................................................................................59
Spatial Point Pattern ..............................................................................59
Quadrant Count Method ....................................................................63
Linear Correlation Statistic.......................................................................63
Case Study................................................................................................64
Statistical Example..................................................................................65
Spatial Correlation Statistics ....................................................................65
Moran's $I$ and Geary's $C$....................................................................66
Cross-Correlation Statistic......................................................................67

Inverse Distance Weighting (IDW)..........................................................67
Statistical Example....................................................................................69
    1. Develop Inverse Distance Weighting........................................69
    2. Develop Moran's *I* ........................................................................69
    3. Develop Geary's *C* .......................................................................71
    4. Develop Bi-Moran's *I*...................................................................73
References ...................................................................................................75

**4. Geospatial Analysis and Modeling–Mapping**................................79
Stepwise Regression..................................................................................79
    Statistical Example ...............................................................................80
Ordinary Least Squares (OLS).................................................................81
Variogram and Kriging..............................................................................83
    Ordinary Kriging...................................................................................85
    Simple Kriging........................................................................................86
    Universal Kriging...................................................................................87
    Developing Variogram Model and Kriging to Predict Plant
    Diversity at GSENM, Utah....................................................................87
        Model Cross-Validation..................................................................91
    Spatial Autoregressive (SAR) ...............................................................91
    Statistical Example ................................................................................92
        Using Spatial AR Model (without Regression) ..........................94
        Using Spatial AR Model (with Regression, OLS Model)
        Using R or S-Plus............................................................................94
        Example on How to Develop Plot of Standard Normal
        Distribution.....................................................................................95
        Analysis of Residuals for Plant Species Richness
        (gsenmplant) Data...........................................................................95
        Weighted SAR Model......................................................................96
Binary Classification Tree (BCTs) .........................................................97
Cokriging ....................................................................................................100
Geospatial Models for Presence and Absence Data ..............................104
    GARP Model............................................................................................105
    Maxent Model.........................................................................................106
    Logistic Regression ...............................................................................106
    Classification and Regression Tree (CART) ......................................107
    Envelope Model......................................................................................108
References .....................................................................................................108

**5. R Statistical Package** ...........................................................................115
Overview of R Statistics (R).....................................................................116
    What Is R? ................................................................................................116
    Strengths of R/S .....................................................................................116
    The R Environment.................................................................................117

Scripts .................................................................................................... 118
Working with R on Your COMPUTER .............................................. 118
Begin to Use R .................................................................................... 118
Statistical Analysis Examples Using R .................................................. 119
Common Statistics ............................................................................... 119
Common Graphics ............................................................................... 119
Common Programming ....................................................................... 120
Create and Examine a Logical Vector ............................................... 121
Working on Graphical Display of Data (Data Distributions) ........... 121
Develop a Histogram ........................................................................... 122
Data Comparison between the Data and an
Expected Normal Distribution ........................................................... 122
More Statistical Analysis .................................................................... 124
Reading New Variable (Enter new data set, WEIGHT) ..................... 124
Plotting Weight and Height ................................................................ 126
Test of Association .............................................................................. 126
Some Basic Regression Analysis ........................................................ 127
Case Study ............................................................................................... 128
Test for Spatial Autocorrelation Using Moran's *I* ............................ 131
Test for Spatial Autocorrelation Using Geary's C ............................ 132
Test for Spatial Cross-Correlation Using Bi-Moran's *I* .................... 133
Trend Surface Analysis .......................................................................... 134
Test for Spatial Autocorrelation of the Residuals ............................ 136
Test for Moran's *I* for Residuals ........................................................ 137
Using Spatial AR Model without Regression ................................... 138
Using Spatial AR with Regression (Using All Independent
Variables as with OLS Model) ........................................................... 138
Analysis of Residuals ......................................................................... 140
Develop Variogram Model (Modeling Fine Scale Variability) ......... 140
Plotting Variogram Model .................................................................. 143
References ................................................................................................ 143

6. **Working with Geospatial Information Data** ....................................... 145
Exercise 1: Working with Remotely Sensed Data ................................. 145
Exercise 2: Derived Remote Sensing Data and
Digital Elevation Model (DEM) .............................................................. 145
Deriving Slope and Aspect from DEM Data ..................................... 147
Resample GRID .................................................................................... 147
Exercise 3: Geospatial Information Data Extraction ............................. 148
Deriving SLOPE and ASPECT from DEM Data (ELEVATION) ....... 149
Resample GRID .................................................................................... 149
Select Area of Interest (Study Site) .................................................... 150
Data Extraction ................................................................................... 150
Steps for Converting the Geospatial Model to a
Thematic Map Product ........................................................................ 152

Working with Vegetation Indices and Tasseled
Cap Transformation .......................................................................... 154
    Vegetation Indices ........................................................... 154
    Tasseled Cap ..................................................................... 155
Develop Thematic Layer in ARCVIEW or ARCMAP ........................ 157
    VIEWS (Working Only with ARCVIEW) ........................................ 157
Map Layout ...................................................................................... 159
References ........................................................................................ 160

**Index** .................................................................................................. 163

# *Preface*

Geospatial information modeling and thematic mapping has become an important tool for investigation and management of natural resources, environment, and sustainability of our living ecosystem at the landscape scale. In writing this book, I hope to share the lessons from my education and working expertise of 20 years in research applications at Colorado State University and other institutes. Further, I have benefited from the mentorship and observation of my colleagues in teaching and research from many different disciplines and institutions. This book is based on a workshop I developed during fall 2006 on integration of geospatial information and spatial statistics for natural resources and environmental applications. I received much support from the people who participated in my workshop and friends, and I was encouraged to use that material as a foundation for this book. There is a need for such a book that can be used by people of different backgrounds and without advanced knowledge of geospatial information sciences such as remote sensing and geographic information systems (GIS), and of quantitative methods including biometry, statistics, sampling methods and design, and geostatistics. This book will attempt to review the types and applications of these geospatial information data (e.g., remote sensing, GIS, and global positioning systems [GPS]) as well as their integration into landscape scale geospatial statistical models and thematic mapping natural resources, and ecological and environmental applications.

The book has different focuses and reference citations. Following is a general description of the basic content related to each chapter to accomplish the goal of this book.

## Principles and Applications

- Chapter 1—Geospatial information sciences and technology in an easy-to-follow format to integrate with geospatial modeling and mapping applications.
- Chapter 2—Collection of data in terms of types, representation, and sources of errors; sampling methods and applications; probability sampling and nonrandom sampling; sampling designs and double sampling and mapping accuracy; plot design and criteria; characteristics of different plot shapes; and plot size.
- Chapter 3—Spatial pattern and correlation statistics, scale and spatial pattern, spatial sampling, errors in spatial analysis, spatial variability and method of predication, spatial pattern, linear correlation and

spatial correlation statistics (Moran's *I*, Geary's *C*, cross-correlation statistics (bi-Moran's *I*), and inverse distance weighting.

- Chapter 4—Geospatial analysis and modeling, thematic mapping, stepwise regression, ordinary least squares (OLS), variogram, kriging, and spatial autoregression, binary classification trees, cokriging, geospatial models for presence and absence data.

- Chapter 5—Topics related to the R Statistical Package are introduced. These topics cover the functions of the software, how to install it on a computer, working on statistical analysis with examples, working on case study, and steps to develop a geospatial statistical model (trend surface analysis) and thematic maps.

- Chapter 6—Integration of geospatial modeling and mapping applications using laboratory exercises. Topics relate to linking field data, integration using remotely sensed data, GIS, and analysis using geospatial modeling and mapping approaches.

My hope is that the reader of this book will be able to analyze geospatially explicit data and to integrate the geospatial information sciences and field data with geospatial statistical modeling and mapping. Also, the reader will be aware of the limitations of geospatial modeling and mapping tools, and will have the opportunity to apply this knowledge. There are many books available for advanced knowledge in the area of remote sensing, GIS, spatial statistics, and I am hoping this book will be very useful to your needs no matter your level of education.

# About the Author

**Dr. Mohammed A. Kalkhan** has over 20 years experience in research and teaching at Colorado State University in Fort Collins, Colorado. As a member of the Natural Resource Ecology Laboratory (NREL), he has also served as an affiliate faculty in the Department of Forest, Rangeland, and Watershed Stewardship, and as an advisor for the Interdisciplinary Graduate Certificate in Geospatial Science, Graduate Degree Program in Ecology (GDPE), The School of Global Environmental Sustainability (SOGES), and Department of Earth Resources (currently the Department of Geosciences) at Colorado State University (CSU). Dr. Kalkhan received his BSc in Forestry (minor in General Agricultural Sciences, 1973) and MSc in Forest Mensuration (1980) from the College of Agriculture and Forestry, the University of Mosul, Iraq. He received his PhD in forest biometrics—remote sensing applications from the Department of Forest Sciences at Colorado State University, USA, in 1994. From 1975 to 1982, he was a lecturer in the Department of Forestry, College of Agriculture and Forestry, University of Mosul, Iraq. In 1994, he joined the Natural Resource Ecology Laboratory. He has served on a number of Program Planning Committees including Monitoring Science and Technology Symposium: Unifying Knowledge for Sustainability in the Western Hemisphere, Sponsored by the USDA Forest Service and EPA; NASA-USGS Invasive Species Tasks; USGS-NPS Mapping, NASA, USDA-Agriculture Research Service (ARS) on a research program to develop predictive spatial models and maps for leafy spurge at Theodore Roosevelt National Park, North Dakota using hyperspectral imaging from NASA EO-1 Hyperion (space), NASA AVIRIS (high altitude aircraft), ARS (CASI- low altitude aircraft); and The First Conference on Fire, Fuels, Treatments and Ecological Restoration: Proper Place, Appropriate Time. Dr. Kalkhan is also a member of the Consortium for Advancing the Monitoring of Ecosystem Sustainability in America (CAMESA).

Dr. Kalkhan's main interests are in the integration of field data, GIS, and remote sensing with geospatial statistics to understand landscape parameters through the use of a complex model with thematic mapping approaches, including sampling methods and designs, biometrics, determination of uncertainty and mapping accuracy assessment. Research applications have included wildfire ecology, wetlands analysis, land use and land cover change, environmental sustainability, ecological forecasting—invasive species, and plant diversity studies. He is also interested in applications for the development of national capacity and technology transfer in developing regions. His research has been aimed at developing a better understanding of landscape-scale ecosystems at any level and to develop better tools for ecological–environmental forecasting. The challenge is to develop new tools based on geospatial information and mathematics–statistics to forecast landscape-scale characteristics.

# 1

## Geospatial Information Technology

Remote sensing systems can only offer the following relationship between spatial and spectral resolution: A high spatial resolution is associated with a low spectral resolution and vice versa. That means that a system with a high spectral resolution can only offer a medium or low spatial resolution. Therefore, it is either necessary to find compromises between the different resolutions according to the individual application or to utilize alternative methods of data acquisition.

To explore and explain how one can extrapolate field data (i.e., vegetation, soil, and environmental) across a landscape scale using geospatial information auxiliary data such as remotely sensed data (e.g., Landsat Multispectral Scanner (MSS) and Thematic Mapper (TM5), Landsat TM-7 Enhanced Thematic Mapper plus (ETM+)), IKONS, and SPOT), geographic information systems (GIS) raster data, and the digital elevation model (DEM) to describe the spatial variability observed in the field data. In order to develop these geospatial models, it is important that we are able to extract the information we need from the remotely sensed imagery, GIS, and global positioning systems (GPS), then combine this information with field data to do the modeling of interest (Figure 1.1). Once we have completed the modeling phase, the last thing that we need to do is to reconstruct the spatial model in ARCINFO, ARCVIEW, or ARCGIS for display. The following are some useful information on geospatial information data and procedures that will help us accomplish this goal.

### Remotely Sensed Data

The first remote sensing satellite was launched on July 23, 1972, and was called ERTS-1 (Earth Resource Technology Satellite). It was an experimental system designed to test the feasibility of collecting Earth resource information (e.g., land cover, land use, Earth environments). On January 22, 1975, the National Aeronautic and Space Administration (NASA) renamed the ERTS program to Landsat, distinguishing it from the Seasat oceanographic satellite launched June 26, 1978. The following table describes the history of the Landsat program.

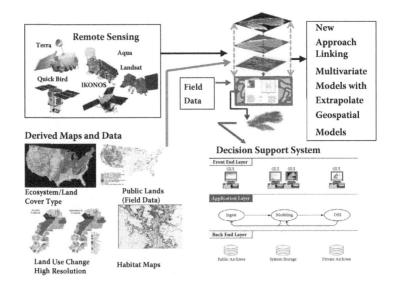

**FIGURE 1.1**
**(See color insert)** Multiple steps that can be used to integrate geospatial information sciences, geospatial forecasting systems, and mapping with decision support system for natural resources management, ecological forecasting, and environmental sustainability applications.

| Satellite | Launched | Date Retired | Altitude | Earth Revisit* | Sensor Types |
|---|---|---|---|---|---|
| Landsat-1 | 7/23/1972 | 1/6/1978 | 917 km | 18 days | MSS[1], RBV[2] |
| Landsat-2 | 1/22/1975 | 2/25/1982 | 917 km | 18 days | MSS, RBV |
| Landsat-3 | 3/5/1978 | 3/31/1983 | 971 km | 18 days | MSS, RBV |
| Landsat-4 | 7/16/1982 | 7/16/1993 | 705 km | 16 days | MSS, TM[3] |
| Landsat-5 | 3/1/1984 | Continuing | 705 km | 16 days | MSS, TM |
| Landsat-6 | 10/5/1993 | Did not achieve orbit | | | |
| Landsat-7 | April 1999 | Continuing | 705 km | 16 day | ETM+[4] |

\* Orbital tracks.
[1] MSS, Multispectral Scanner (takes measurement in four bands of the electromagnetic spectrum).
[2] RBV, Return Beam Videocon.
[3] TM, Thematic Mapper (takes measurement in seven bands of the electromagnetic spectrum).
[4] The ETM+ instrument is the Enhanced Thematic Mapper Plus with eight bands of multispectral scanning radiometer capable of providing high-resolution imaging information of Earth's surface. On May 31, 2003 the Scan Line Corrector (SLC) in the ETM+ instrument failed.

The new Landsat-7 detects spectral information in eight bands: the panchromatic has a 15 m × 15 m resolution; the six visible, near and short-wave infrared bands have a resolution of 30 m × 30 m; the thermal band has a resolution of 60 m × 60 m. For a detailed history of the Landsat program, refer to the Landsat Data User Notes published by the EROS Data Center (NOAA, 1975–1985) and EOSAT, Inc. (EOSAT, 1986–present). Description of Landsat MSS and TM sensor system characteristics are given next.

| Landsat Multispectral Scanner (MSS) | | Landsat Thematic Mapper (TM) | |
|---|---|---|---|
| Band | Micrometer (µm) | Band | Micrometer (µm) |
| 4[a] | 0.5–0.6 (green) | 1 | 0.45–0.52 (blue) |
| 5 | 0.6–0.7 (red) | 2 | 0.52–0.60 (green) |
| 6 | 0.7–0.8 (reflective infrared) | 3 | 0.63–0.69 (red) |
| 7 | 0.8–1.1 (reflective infrared) | 4 | 0.76–0.90 (reflective infrared) |
| 8[b] | 10.4–12.6 (thermal infrared) | 5 | 1.55–1.75 (mid-infrared) |
| | | 6 | 10.4–12.5 (thermal infrared) |
| | | 7 | 2.08–2.35 (mid-infrared) |

[a] MSS band 4, 5, 6, and 7 were renumbered bands 1, 2, 3, and 4 on Landsat-4 and 5.
[b] MSS band 8 was present only on Landsat-3.

### Instantaneous Field of View (IFOV) at Nadir (Resolution on the Ground)

Instantaneous field of view (IFOV) for MSS bands 4, 5, 6, and 7 is 79 m × 79 m spatial resolution and band 8 is 240 m × 240 m spatial resolution. Each MSS scene represents a 185 km × 178 km (32,930 km$^2$) or 115 miles × 111 miles (12,765 mi$^2$). However, the IFOV for Landsat TM bands 1 to 5, and 7 is 30 m × 30 m (spatial resolution), while band 6 has 120 m × 120 m resolution. The area covered is similar to Landsat MSS. A typical scene contains approximately 2340 scan lines with about 3240 pixels per line or about 7,581,600 pixels per channel. Both sensors cover large areas of the landscape and is more cost efficient than alternative methods of data collection. However, problems persist in using Landsat for thematic mapping classification due to misclassification errors and cost of field surveys to accomplish acceptable accuracy assessment.

Landsat MSS bandwidths were selected based on their utility for general vegetation inventories (quantitative and qualitative analysis) and geological studies. However, the Landsat TM bands were chosen after years of analysis for their value in water penetration; discriminating vegetation types and vigor; plant and soil moisture measurements; differentiation of clods, snow, and ice; and identification of hydrothermal alteration of certain rocks (Jensen 1996). The characteristics of the Landsat TM spectral bands are described and cited by Jensen (1996, p. 40). See Jensen (1996) for more detailed information about Landsat MSS and TM.

Band 1 (0.45 to 0.52 μm; blue)—This band provides increased penetration of water bodies; supports the analysis of land-use, soil, and vegetation characteristics.

Band 2 (0.52 to 0.60 μm; green)—The band spans the region between the blue and red chlorophyll absorption bands and therefore corresponding to the green reflectance of healthy vegetation.

Band 3 (0.63 to 0.69 μm; red)—This is the red chlorophyll absorption band of healthy vegetation and represents one of the most important vegetation discrimination. It is also useful for soil-boundary and geological-boundary delineations.

Band 4 (0.76 to 90 μm; reflective infrared)—This band is especially responsive to the amount of vegetation biomass present in a scene. It is useful for crop identification and emphasizes soil–crop and land–water contrast.

Band 5 (1.55 to 1.75 μm; mid-infrared)—This band is sensitive to amount of water in plants. This information is useful for crop drought studies and plant vigor investigation. This band also used to discriminate between clouds, snow, and ice, which is very important in hydrologic studies.

Band 6 (10.4 to 12.5 μm; thermal infrared)—This band is useful for geothermal activity, thermal inertia for geological investigations, vegetation classification, vegetation stress analysis, and soil moisture studies. Also, it can be used for topographic studies in mountain terrain.

Band 7 (2.08 to 2.35 μm; mid-infrared)—It is useful for discrimination of geological rock information and identification zones of hydrothermal alteration in rocks.

## IKONOS

IKONOS is derived from the Greek word for "image." The IKONOS satellite is the world's first commercial satellite to collect black-and-white images with 1 m resolution and multispectral imagery with 4 m resolution. The IKONOS satellite produces color imagery with a resolution of .82 m (82 cm or about 32 in.) to 3.2 m with positional and relative accuracy suitable for virtually all mapping requirements. Its mapping, intelligence analysis, and feature extraction capabilities are among the best of any commercial imagery system in its class.

IKONOS standard products include 1 m black-and-white, 4 m multispectral (all bands), 1 m color (true color, false color, or 4 band), and a 1 m and

4 m data bundle. IKONOS image data is available in easy-to-use 8 bit or full dynamic range 11 bit format. Most of the sensor data are in application uses: urban city planning, specific site locations, forest health and fire hazards, limited vegetation study, national security, military mapping, air and marine transportation, and by regional and local governments.

### Sensor Characteristics

The IKONOS satellite weighs about 1600 pounds. It orbits the Earth every 98 minutes at an altitude of approximately 680 km or 423 miles. IKONOS was launched into a sun-synchronous orbit, passing a given longitude at about the same local time (10:30 a.m.) daily. IKONOS can produce 1 m imagery of the same geography every 3 days.

### IKONOS Specifications

Spatial resolution: 0.82 m × 3.2 m; Spectral range: 526–929 nm
Bands:
Blue: 445–516 nm
Green: 506–595 nm
Red: 632–698 nm
Near-IR: 757–853 nm
Swath width: 11.3 km
Off-nadir imaging: Up to 60 degrees; Dynamic range: 11 bits per pixel
Mission life expected: More than 8.3 years; Revisit time: approximately 3 days
Orbital altitude: 681 km; Nodal crossing: 10:30 a.m.

### ORBIMAGE (GeoEye)

ORBIMAGE (GeoEye) is a leading global provider of satellite-delivered Earth imagery services with a planned constellation of five digital remote sensing satellites. It currently operates the OrbView-1 atmospheric imaging satellite (launched in 1995); the OrbView-2 ocean and land multispectral imaging satellite (launched in 1997); and an integrated image receiving, processing, and distribution system with over 20 current and planned regional ground stations around the world. As the foundation upon which GeoEye built the SeaStar[SM] Fisheries Information Service, OrbView-2 provides imagery for maps used by commercial vessels to detect favorable oceanographic fishing conditions. The satellite also provides broad-area coverage in 2800 km wide swaths, which are routinely used in naval operations, environmental monitoring, and global crop-assessment applications.

### *OrbView-2 Specifications*

Spatial Resolution: 1.13 km LAC/HRPT and 4.5 km GAC; Spectral range: 0.402–0.888 nm

Bands

Violet: 402–422 nm; Violet-blue: 433-453 nm

Blue: 480–500 nm

Green: 500–520 nm; Green: 545–565 nm

Red: 660–680 nm

Near-IR: 745–785 nm

Near-IR: 845–885 nm

Orbit type: Sun synchronous at 705 km

Nodal Crossing Noon: +20 min, descending; Orbital period: 99 minutes

Swath width: 2800 km LAC/HRPT (58.3 degrees); 1500 km GAC (45 degrees)

Revisit time: 1 day; Calibration: On-board solar diffuser calibration updates reference

Lunar calibration: Monthly maneuver; Digitization: 10 bits

Sun glint avoidance: ±20 degree sensor fore-aft tilt

Mission life: More than 12 years

### *OrbView-3 Specifications*

OrbView-3, the first of the high-resolution (1 m panchromatic and 4 m multi-spectral) optical imaging satellites, was launched in June 26, 2003, followed by the launch in late 2000 of OrbView-4, which include a hyperspectral imagery capability. In addition, ORBIMAGE is a nonexclusive distributor of Russian SPIN-2 imagery and holds the exclusive, worldwide distribution rights for the Canadian RadarSat-2 satellite.

Orbit: 470 km inclined at 97°/470 km, sun synchronous

Stabilization: 3-axis control

Mass: 304 kg (670 lbs)

Swath width: 8 km

Mission life: 5 years, mission is completed

### QuickBird

The QuickBird satellite is the first in a constellation of spacecraft owned and operated by Digital Globe (Longmont, Colorado), which offers highly accurate, commercial high-resolution imagery of Earth. QuickBird's global collection of panchromatic and multispectral imagery is designed to support applications ranging from map publishing to land and asset management to insurance risk assessment. Using a state-of-the-art BGIS 2000 sensor, QuickBird collects image data to 0.61 m pixel resolution degree of detail. This satellite is an excellent source of environmental data useful for analyses of changes in land usage–land cover, agricultural and sustainability, forest structure, ecosystem, climates and global change, and environmental forecasting. QuickBird's imaging capabilities can be applied to a host of industries, including oil and gas exploration and production, engineering and construction, defense homeland security and military application, and environmental studies.

### *QuickBird Satellite Sensor Characteristics*

Launch date: October 18, 2001; Launch vehicle: Boeing Delta II

Orbit altitude: 450 km; Orbit inclination: 97.2°, sun synchronous

Speed: 7.1 km/sec (25,560 km/hour)

Equator crossing time: 10:30 a.m. (descending node); Orbit time: 93.5 minutes

Revisit time: 1–3.5 days, depending on latitude (30° off-nadir)

Swath width: 16.5 km × 16.5 km at nadir

Metric accuracy: 23 m horizontal (CE90%)

Digitization: 11 bits

Band spectral resolution (range):

Panchromatic mode: 61 cm (nadir) to 72 cm (25° off-nadir)

Multispectral mode: 2.44 m (nadir) to 2.88 m (25° off-nadir)

Bands spectral resolution:

Pan: 450–900 nm

Blue: 450–520 nm

Green: 520–600 nm

Red: 630–690 nm

Near-IR: 760–900 nm

### The SPOT (System Probatori d'Observation de la Terre)

SPOT Image is the commercial operator of three European SPOT satellites. SPOT image has become the world's leading supplier of satellite-based geographic information, with its international distribution network consisting of four subsidiaries (United States, Australia, Singapore, and China), eighty distributors, and a comprehensive network of twenty-three direct-receiving stations. Its principal shareholders are the Center National d'Etudes Spatiales

(CNES; the French space agency) and major European private companies such as Aerospatiale-Marta group, Alcatel Space Industries, and the Swedish Space Corporation.

SPOT multispectral imagery is used in vegetation mapping because although the spectral resolution is not as good as Landsat (narrow spectral resolution of the band), the spatial resolution is higher. SPOT is also used for soil erosion and urban mapping, forest resources mapping, and planning. SPOT multispectral imagery is used mostly for geological, soil, and urban studies. Due to limited multispectral atmospheric windows (narrow 3-4 bands), the use of SPOT data in vegetation studies has been limited. An advantage of using SPOT data is the ability of stereoscopic viewing, which can be used to produce topographic and plan metric thematic maps. The major problem using Landsat MSS, TM, TM+, MODIS, SPOT, or other remotely sensed data is the influence of cloud cover and the restriction to daytime use. This makes it hard in some parts of the globe to require free cloud imagery. The alternative is to use radar imagery in some situations, if there is a need.

SPOT-1 satellite was developed by the CNES and launched February 21, 1986, from French Guiana and was the first Earth resource satellite to utilize a linear array sensor scanning by means of push broom scanning techniques. SPOT-1 is a high resolution visible sensor system (HRVSS) has a spatial resolution of 10 m × 10 m (panchromatic mode) and 20 m × 20 m (multispectral mode), and provides several other innovations in remote sensor system design. In addition, the multispectral HRVSS consists of three or four bands, and is designed to operate over the 500–1750 nm range of the spectrum. The multispectral sensor system on the SPOT satellites is pointable. This means that off-nadir image capturing is possible to enable stereoscopic image capture.

SPOT-2, which carries the XS multispectral scanner, and SPOT-3, with identical payloads, were launched in January 20, 1990, and September 25, 1993, respectively. (SPOT-3 failed approximately a year after launch.) SPOT-2 scans using three bands picks up energy in the green, red, and near-infrared part of the spectrum; whereas SPOT-4, which carries the XI multispectral scanner, was launched in March 1998 and operates in a sun-synchronous, near-polar orbit (inclination of 98.7°) at an altitude of 832 km. SPOT-4 offers a unique worldwide imaging capability. SPOT 5 was successfully placed into orbit by an Ariane 4 from the Guiana Space Centre in Kourou during the night of May 3–4, 2002. The SPOT-5 Earth observation satellite has improved capabilities (2.5 m resolution with 60 km field of view, along-track stereoscopic viewing capacity) with revisit time 2–3 days depending on latitude (orbital altitude = 822 km).

### *SPOT-5 Satellite Sensor Characteristics*

Panchromatic band resolution: 2.5 m from 2 × 5 m scenes, and 5 m (nadir)

Multispectral band resolution: 10 m (nadir); SWI: 20 m (nadir)

In general, the SPOT sensor system characteristics are:

|  | Multispectral Mode (μm) | Panchromatic Mode (μm) |
|---|---|---|
| Band 1 | 0.50–0.59 | 0.48–0.71 |
| Band 2 | 0.61–0.68 | |
| Band 3 | 0.79–0.89 | |
| Shortwave IR | 1580–1750 nm | |
| Number of pixels per line | 3000 | 6000 |
| Ground swath width at nadir | 60 km | 60 km |

## MODIS (Moderate Resolution Imaging Spectroradiometer)

Global change and ecological–environmental forecasting research investigates the underlying processes of change and their manifestation, the impacts and the prediction of change. Monitoring these changes provides an important underpinning to both global change research and resource management. The Moderate Resolution Imaging Spectroradiometer (MODIS) is providing systematic measurements in support of NASA's Earth Science Enterprise. High-quality, consistent, and well-calibrated satellite measurements are needed to detect and monitor changes and trends in these variables. Developing the next-generation data sets for global change research is the challenge given to the MODIS Science Team. For more information on MODIS, visit the Web sites http://terra.nasa.gov/.

MODIS, which was launched on the Terra platform December 18, 1999, is designed (wide spectral range, moderate spatial resolution [250 m–1 km], and near daily global coverage) to observe and monitor the surface of the Earth. MODIS is one of the key instruments for NASA's Earth Observing System (EOS), built by Raytheon Santa Barbara Remote Sensing (SBRS). In addition, MODIS Proto Flight Model (PFM) was launched on board the Terra spacecraft on December 28, 1999 (first light on February 24, 2000). The second MODIS was launched on the Aqua platform on May 4, 2002 (first flight on June 24, 2002). Also, MODIS Flight Model 1 (FM1) on the Aqua spacecraft was launched with 36 reflective spectral bands as:

20 reflective solar bands (RSB): 0.4–2.2 microns

16 thermal emissive bands (TEB): 3.5–14.5 microns

In addition to advising on the instrument specifications and design, the MODIS Land Group (MODLAND) is tasked with developing algorithms, and generating and validating the data products. These higher order data products have been designed to remove the burden of certain common types of data processing from the user community and meet the more general needs of global-to-regional monitoring, modeling, and assessment.

### MODIS Overview

36 spectral bands (490 detectors) cover wavelength range from 0.4 to 14.5 mm

Spatial resolution at nadir: 250 m (2 bands), 500 m (5 bands), and 1000 m

4 FPAs: VIS, NIR, SMIR, LWIR (see Figure 1.2)

On-board calibrators: SD/SDSM, SRCA, and BB (plus space view)

12 bit (0-4095) dynamic range

2-sided Paddle Wheel Scan Mirror scans 2330 km swath in 1.47 sec

Day data rate = 10.6 Mbps; night data rate = 3.3 Mbps (100% duty cycle, 50% day and 50% night)

### Technical Specifications of MODIS

(Further information can be found at www.raytheon.com.)

Orbit: 705 km, 10:30 a.m. descending node or 1:30 p.m. ascending node, sun synchronous, near-polar, circular

Scan rate: 20.3 rpm, cross-track

Scan dimensions: 2330 km (cross-track) by 10 km (along track at nadir)

Telescope: 17.78 cm dia. off-axis, a focal (collimated), with intermediate field stop

Size: 1.0 × 1.6 × 1.0 m

Mass: 240 kg

Power: 150 W (orbital average)

Data rate: 11 Mbps (peak daytime)

Quantization: 12 bits

Spatial resolution: 250 m (bands 1–2), 500 m (bands 3–7), 1000 m (bands 8–36)

Mission life expected: 5 years

### MODIS Land Products

There are many MODIS products. Following are some examples:

Energy Balance Product Suite
- Surface Reflectance
- Land Surface Temperature, Emissivity
- BRDF/Albedo
- Snow/Sea-Ice Cover

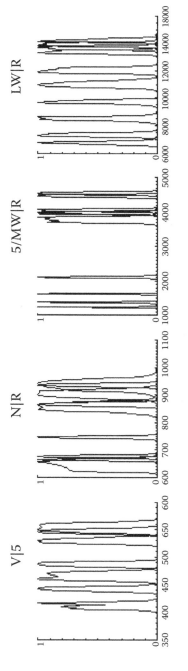

**FIGURE 1.2**
MODIS Spectral wavelength information (see NASA http://terra.nasa.gov/ and http://eos-pm.gsfc.nasa.gov/; USGS http://USGS. gov is a general Web site used for more information about MODIS).

Vegetation Parameters Suite
- Vegetation Indices
- LAI/FPAR
- GPP/NPP

Land Cover/Land Use Suite
- Land Cover/Vegetation Dynamics
- Vegetation Continuous Fields
- Vegetation Cover Change
- Fire and Burned Area

MODIS Resolution Spectral Bands

| Primary Use | Band | Bandwidth[1] | Spectral Radiance[2] | Require SNR[3] NE$\Delta$T(K)[4] |
|---|---|---|---|---|
| Land/Cloud Boundaries | 1 | 620–670 | 21.8 | 128 |
| | 2 | 841–876 | 24.7 | 201 |
| Land/Cloud Properties | 3 | 459–479 | 35.3 | 243 |
| | 4 | 545–565 | 29.0 | 228 |
| | 5 | 1230–1250 | 5.4 | 74 |
| | 6 | 1628–1652 | 7.3 | 275 |
| | 7 | 2105–2155 | 1.0 | 110 |
| Ocean Color/Phytoplankton/ Biogeochemistry | 8 | 405–420 | 44.9 | 880 |
| | 9 | 438–448 | 41.9 | 838 |
| | 10 | 483–493 | 32.1 | 802 |
| | 11 | 526–536 | 27.9 | 754 |
| | 12 | 546–556 | 21.0 | 750 |
| | 13 | 662–672 | 9.5 | 910 |
| | 14 | 673–683 | 8.7 | 1087 |
| | 15 | 743–753 | 10.2 | 586 |
| | 16 | 862–877 | 6.2 | 516 |
| Atmospheric Water Vapor | 17 | 890–920 | 10.0 | 167 |
| | 18 | 931–941 | 3.6 | 57 |
| | 19 | 915–965 | 15.0 | 250 |
| Surface/Cloud Temperature | 20 | 3.66–3.840 | 0.45 | 0.05 |
| | 21 | 3.929–3.989 | 2.38 | 2.00 |
| | 22 | 3.929–3.989 | 0.67 | 0.07 |
| | 23 | 4.020–4.080 | 0.79 | 0.07 |
| Atmospheric Temperature | 24 | 4.433–4.498 | 0.17 | 0.25 |
| | 25 | 4.482–4.549 | 0.59 | 0.25 |
| Cirrus Clouds | 26 | 1.360–1.390 | 6.00 | 1503 |
| Water Vapor | 27 | 6.535–6.895 | 1.16 | 0.25 |
| | 28 | 7.175–7.475 | 2.18 | 0.25 |

MODIS Resolution Spectral Bands (Continued)

| Primary Use | Band | Bandwidth[1] | Spectral Radiance[2] | Require SNR[3] NEΔT(K)[4] |
|---|---|---|---|---|
| | 29 | 8.400–8.700 | 9.58 | 0.05 |
| Ozone | 30 | 9.580–9.880 | 3.69 | 0.25 |
| Surface/Cloud Temperature | 31 | 10.780–11.280 | 9.55 | 0.05 |
| | 32 | 11.770–12.270 | 8.94 | 0.05 |
| Cloud Top Altitude | 33 | 13.185–13.485 | 4.52 | 0.25 |
| | 34 | 13.485–13.785 | 3.76 | 0.25 |
| | 35 | 13.785–14.085 | 3.11 | 0.25 |
| | 36 | 14.085–14.385 | 2.08 | 0.35 |

[1]  Bands 1 to 19, nm; Bands 20 to 36, μm
[2]  (W/m²-μm-sr)
[3]  SNR, signal-to-noise ratio
[4]  NEΔT(K), noise-equivalent temperature difference
[3 & 4]  Means performance goal is 30–40% better than required
   The above information was cited from: www.raytheon.com and http://terra.nasa.gov/.

## ASTER (Advanced Spaceborne Thermal Emission and Reflection radiometer)

The first Earth Observing System (EOS) satellite called Terra (previously AM-1) was launched on December 18, 1999, from Vandenberg Air Force Base in California. Terra will fly in a sun-synchronous polar orbit, crossing the equator in the morning at 10:30. ASTER is one of the five state-of-the-art instrument sensor systems on board Terra with a unique combination of wide spectral coverage and high spatial resolution in the visible near-infrared through shortwave infrared to the thermal infrared regions. It was built by a consortium of Japanese government, industry, and research groups. ASTER data is expected to contribute to a wide array of global change-related application areas including vegetation and ecosystem dynamics, hazard monitoring, geology and soils, land surface climatology, hydrology, and land-cover change.

### ASTER Uniqueness

- The visible near-infrared (VNIR) telescope's backward viewing band for high-resolution along-track stereoscopic observation.
- Multispectral thermal infrared data of high spatial resolution (8 to 12 μm window region, globally).
- Highest spatial resolution surface spectral reflectance, temperature, and emissivity data within the Terra instrument suite.
- Capability to schedule on-demand data acquisition requests.

### History of ASTER

- 1981—NASA commissioned an in-house study to determine requirements for a polar-orbiting platform to provide Earth science observations, resulting in System Z, now the EOS. Among the early EOS designs, one of the strawman instruments was JPL's Thermal Infrared Multispectral Scanner (TIMS) as follow-up to the Airborne TIMS.

- 1988—TIMS' instrument design was refined and proposed as the Thermal Infrared Ground Emission Radiometer (TIGER). TIGER had 2 components: TIMS (14 channels in the 3–5µ and 8–15µ regions) and Thermal Infrared Profiling System (TIPS).

- Around the same time, Japan's MITI (Ministry of International Trade & Industry) offered to provide the Intermediate Thermal Infrared Radiometer (ITIR) which measured radiances in 11 bands in NIR, SWIR, and TIR regions. NASA accepted this design and asked the TIGER team to implement its design advances by influencing the Japanese design of ITIR. ITIR was later redesigned to include 14 channels in the visible near-infrared, shortwave infrared, and thermal infrared regions, and renamed ASTER (JPL, 1994).

### Organizational Framework of ASTER

There are a number of entities in the United States and Japan that are involved in the development and production of ASTER data and data products. These include, for instance, the satellite sensor systems and its operations, data reception, processing, management, data product development, quality assurance, distribution, archival, and storage. The following tables outline these entities.

ASTER Sensor Systems: Baseline Performance Requirements

| Sub-system | Band Number | Spectral Range (µm) | Radiometric Resolution | Absolute Accuracy (σ) | Spatial Resolution | Signal Quantization Levels |
|---|---|---|---|---|---|---|
| VNIR | 1 | 0.52–0.60 | NE Δρ 0.5% | ≤±4% | 15 m | 8 bits |
| | 2 | 0.63–0.69 | | | | |
| | 3N | 0.78–0.86 | | | | |
| | 3B | 0.78–0.86 | | | | |
| SWIR | 4 | 1.600–1.700 | NE Δρ≤ 0.5% | ≤±4% | 30 m | 8 bits |
| | 5 | 2.145–2.185 | NE Δρ≤ 1.3% | | | |
| | 6 | 2.185–2.225 | NE Δρ≤ 1.3% | | | |
| | 7 | 2.235–2.285 | NE Δρ≤ 1.3% | | | |
| | 8 | 2.295–2.365 | NE Δρ≤ 1.0% | | | |
| | 9 | 2.360–2.430 | NE Δρ≤ 1.3% | | | |

ASTER Sensor Systems: Baseline Performance Requirements (Continued)

| Sub-system | Band Number | Spectral Range (μm) | Radiometric Resolution | Absolute Accuracy (σ) | Spatial Resolution | Signal Quantization Levels |
|---|---|---|---|---|---|---|
| TIR | 10 | 8.125–8.475 | NE ΔT≤ 0.3% | ≥ 3K (200–240K) | 90 m | 12 bits |
| | 11 | 8.475–8.825 | | ≥ 2K (240–270K) | | |
| | 12 | 8.925–9.275 | | ≥ 1K (270–340K) | | |
| | 13 | 10.25–10.95 | | ≥ 2K (340–370K) | | |
| | 14 | 10.95–11.65 | | | | |

ASTER System Baseline Performance Requirements

| | |
|---|---|
| Swath Width | 60 kms |
| Total Cross-Track Coverage | ± 116 to ± 318 Kkms |
| Stereo Base-to-Height Ratio | 0.6 (along-track) |
| Modulation Transfer Frequency | 0.25 (cross-track) |
| | 0.20 (along-track) |
| Band-to-Band Registration | 0.2 pixels (intra-telescope) |
| | 0.3 pixels (inter-telescope) |
| Duty cycle | 8% (VNIR and SWIR) |
| | 16% (TIR) |
| Peak Data Rate | 89.2 mbps |
| Mass | 406 kgs |
| Peak power | 726 W |

## Active Remotely Sensed Data

### Radar

Radar (radio detection and ranging) is an active sensor that generates its own energy based on signal (beam) return. The wavelengths range from 0.1 to 30 cm, and with different bands or modes (C, K, L, Q, and X). Radar was developed during World War I by the British government for military purposes. The advantage of using radar sensor is the ability to provide information day or night and penetrate of cloud cover. Most applications are focused on tropical forests, soils (i.e., soil moisture), military, geological, and weather studies. However, the launch of European Remote Sensing Satellite (ERS-1) provides high-spatial resolution and work in active microwave instruments (AMI) "SAR Image Mode using 5.3 HGz C-band." The resolution ranges from ≤26.3 × 30 m to ≥45 km with an altitude of 785 km and a swath ranging from 9.6 to 12, 100, and 500 km, and a temporal resolution of 35 days. One of the most common remote sensors used in the 1990s is the Canadian Remote Sensing Satellite called RADARSAT, Synthetic Aperture RADAR (SAR). It operates in seven modes using HH (H = horizontal) polarization and it orbits at an altitude ranging from 793 km to 821 km with a temporal resolution of 4 to 6 days. The image resolution for standard mode is 25 m × 28 m with a swath of 100 km. For more information, see the manual of remote sensing published by American Society of Photogrammetric and Remote Sensing, and other text related to radar or image processing of remotely sensed data.

Remotely sensed data provide unique opportunity for direct or indirect information or measurements about the Earth at different scales and patterns. They provide a full coverage of the landscape, which is desirable for use by researchers, resource management teams, and general publics. Thus, we can integrate remotely sensing data, GIS, and GPS with geospatial statistical for modeling and mapping applications of landscape assessment, forest parameter characteristics, fuel loading and behavior characteristics, wetland ecosystems and their functional parameters, forest health and hazards, plant diversity, invasive species monitoring and forecasting, wildlife habitat, watershed studies, infect diseases, urban planning and growth, and more (see Figures 1.3 and 1.4).

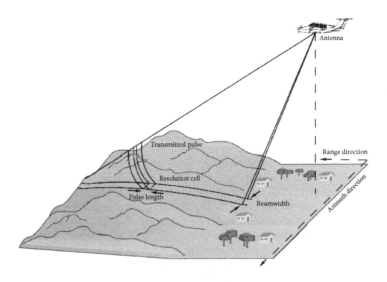

**FIGURE 1.3**
Geometry of a side-looking airborne radar system. (Originally adapted from Lillesand, T. M., and R. W. Kiefer, 1987, *Remote Sensing and Image Interpretation*, 2nd ed., New York, John Wiley & Sons; and cited from http://forsys.cfr.washington.edu/JFSP06/radar_overview.htm.)

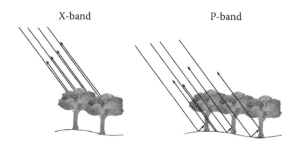

X-band            P-band

**FIGURE 1.4**

Short wavelength X-band radar energy reflects from canopy surface while long wavelength P-band energy penetrates through canopy and reflects from stems and terrain surface. (Originally adapted from Moreira, J., et al., 2001, Surface and ground topography determination in tropical rainforest areas using airborne interferometric SAR, *Photogrammetric Week 01*, D. Fritsch and R. Spiller, Eds., Heidelberg, Herbert Wichmann Verlag.; and cited from http://forsys.cfr.washington.edu/JFSP06/radar_overview.htm.)

## Lidar

Light detection and ranging (lidar) uses the same principle as radar. (This information is from http://www.ghcc.msfc.nasa.gov/sparcle/sparcle_tutorial.htm.) The lidar instrument transmits light out to a target. The transmitted light interacts with and is changed by the target. Some of this light is reflected or scattered back to the instrument where it is analyzed. The change in the properties of the light enables some property of the target to be determined. The time for the light to travel out to the target and back to the lidar is used to determine the range to the target.

There are three basic generic types of lidar:

* Range finders
* DIAL
* Doppler lidars

Range finder lidars are the simplest lidars. They are used to measure the distance from the lidar instrument to a solid or hard target.

A differential absorption lidar (DIAL) is used to measure chemical concentrations (such as ozone, water vapor, pollutants) in the atmosphere. A DIAL uses two different laser wavelengths, which are selected so that one of the wavelengths is absorbed by the molecule of interest while the other wavelength is not. The difference in intensity of the two return signals can be used to deduce the concentration of the molecule being investigated.

A Doppler lidar is used to measure the velocity of a target. When the light transmitted from the lidar hits a target moving towards or away from the lidar, the wavelength of the light reflected/scattered off the target will be changed slightly. This is known as a Doppler shift, hence Doppler lidar.

If the target is moving away from the lidar, the return light will have a longer wavelength (sometimes referred to as a red shift); if moving toward the lidar the return light will be at a shorter wavelength (blue shifted). The target can be either a hard target or an atmospheric target (the atmosphere contains many microscopic dust and aerosol particles that are carried by the wind). These are the targets of interest to us as they are small and light enough to move at the true wind velocity and thus enable a remote measurement of the wind velocity to be made.

### Lidar System Differences

- Coherent (heterodyne) vs. noncoherent (direct) detection
- Pulsed vs. continuous wave (CW)
- Collimated vs. focused
- Monostatic vs. bistatic (biaxial only)
- Coaxial (monostatic only) vs. biaxial
- Doppler vs. non-Doppler
- Ranging vs. nonranging
- Single wavelength vs. DIAL
- Platform: laboratory, ground-mobile, airplane, balloon, ship, spacecraft

### How Does Lidar Work?

The following is a brief description provided by the use of lasers, which has become commonplace, from laser printers to laser surgery. In airborne-laser-mapping lidar, lasers are taken into the sky. Instruments are mounted on a single- or twin-engine plane or a helicopter. Airborne lidar technology uses four major pieces of equipment (see Figure 1.5). These are a laser emitter–receiver scanning unit attached to the aircraft; global positioning system (GPS) units on the aircraft and on the ground; an inertial measurement unit (IMU) attached to the scanner, which measures roll, pitch, and yaw of the aircraft; and a computer to control the system and store data. Several types of airborne lidar systems have been developed; commercial systems commonly used in forestry are discrete-return, small-footprint systems. "Small footprint" means that the laser beam diameter at ground level is typically in the range of 6 inches to 3 feet. The laser scanner on the aircraft sends up to 100,000 pulses of light per second to the ground and measures how long it takes each pulse to reflect back to the unit. These times are used to compute the distance each pulse traveled from scanner to ground. The GPS and IMUs determine the precise location and attitude of the laser scanner as the pulses are emitted, and an exact coordinate is calculated for each point. The laser scanner uses an oscillating mirror or rotating prism (depending on the sensor model), so that the light pulses sweep across a swath of landscape below

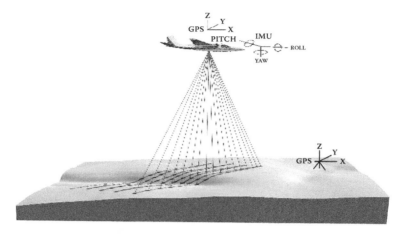

**FIGURE 1.5**
**(See color insert)** Lidar sensor operation. (This figure is property of and cited from http://forsys.cfr.washington.edu/JFSP06/lidar_technology.htm.)

the aircraft. Large areas are surveyed with a series of parallel flight lines. The laser pulses used are safe for people and all living things. Because the system emits its own light, flights can be done day or night, as long as the skies are clear.

Thus, with distance and location information accurately determined, the laser pulses yield direct, 3-D measurements of the ground surface, vegetation, roads, and buildings. Millions of data points are recorded; so many that lidar creates a 3-D data cloud. After the flight, software calculate the final data points by using the location information and laser data. Final results are typically produced in weeks, whereas traditional ground-based mapping methods took months or years. The first acre of a lidar flight is expensive, owing to the costs of the aircraft, equipment, and personnel. But when large areas are covered, the costs can drop to about $1 to $2 per acre. The technology is commercially available through a number of sources.

## Derived Remotely Sensed Data

### Vegetation Indices

A vegetation index is a simple mathematical formula (band ratio) used in remote sensing to estimate the likelihood that vegetation was actively growing at a particular location whenever it was observed. The result of the analysis of remote sensing data with a vegetation index is sometimes also (abusively) called a vegetation index. Typically, a vegetation index permits computing of a single number on the basis of (usually) two or (sometimes) more contemporaneous spectral observations of the target of interest. The proper use and interpretation of this index depends on how it was constructed. Two classes of vegetation indices can be distinguished:

Most indices were empirically constructed in such a way that larger values correspond to higher probabilities of actually finding live green plants at the selected location and time of observation. As the name implies, the result is a nondimensional index or indicator of the presence of vegetation. Empirical indices have often been grossly abused to estimate a wide variety of environmental variables loosely connected to the presence of vegetation.

A new generation of spectral indices has emerged in the second half of the 1990s. These are optimized to estimate a particular environmental variable on the basis of data from a specific instrument. Since they genuinely estimate a geophysical variable, these tools should in fact not be called indices. That label remains largely for historical reasons.

Users of vegetation indices often ask these questions:

What are the advantages of all vegetation indices?
- Simplicity of implementation
- Low computational cost
- Reduction in data volume
- Fast estimate

What are the drawbacks of a classical vegetation index?
- Sensitivity to many geophysical variables not of direct interest or relevance (e.g., atmospheric effects, soil brightness changes with humidity)
- Sensitivity to geometrical conditions of illumination and observation, as well as to the particular anisotropy of the observed system
- Sensitivity to the particular sensor used
- No clear, unique, definite meaning or interpretation
- Not a measurable geophysical product

Note: Values from derive indices are difficult or impossible to compare with great accuracy if they describe locations separated by large distances, locations observed on different days, or locations observed with different sensors.

What benefits derive from an optimal vegetation index?
- The value best estimates a well-defined geophysical variable, which can in principle be measured on the ground.
- The accuracy of the estimate may be estimated.

- Perturbations from known undesirable effects are explicitly minimized
- The quality and performance of optimized indices are fully documented in the refereed literature as well as in technical reports.

Similar products can genuinely be compared across space, time, and sensors.

When can I use a vegetation index?

- A vegetation index may be appropriate when only a rough preliminary indicator of the presence of live green plants, or an approximate estimate of vegetation cover is needed, and this approach will meet those requirements.
- Very large amounts of data need to be processed in a short period of time or at a very limited cost.

When should I not use a vegetation index?

- A vegetation index is inappropriate when a precise quantitative estimation of a property of live green plants is required

How do I choose the best vegetation index for my purpose?

Assuming a vegetation index is the correct approach, the selection of an appropriate index for a particular application depends on your specific needs (what you want to know) and on the observational data available (which sensor can be used). Optimized vegetation indices are designed to best estimate a particular geophysical variable on the basis of data from a specific instrument. Users should thus select the tool that will provide the desired information on the basis of available data.

How accurate is a vegetation index?

The performance of vegetation indices can be evaluated in the same way as any instrument: with the help of a single number, called the signal to noise ratio (or SNR). This numerical value is estimated by dividing the typical range of variation in the signal by the typical amount of noise of variation due to one or more causes. The higher the number, the better. This methodology have been described at length in Leprieur et al. (1994).

Where can I find more about optimized vegetation indices?

The general philosophy behind the design of optimized vegetation indices has been described in Verstraete and Pinty (1996). This approach has been further developed and applied in the context of the exploitation of recent or near future sensors, including:

- SeaWiFS: See Gobron et al. (2002).
- Vegetation: See the Web site of the Preparatory Programmer (http://www.yves-govaerts.com/1/lsp_facosi.php).

- MISR: A publication in under preparation.
- MERIS: See Gobron et al. (1998).

## The Tasseled Cap Transformation

The tasseled cap transformation is derived of remotely sensed data and is the conversion of the readings in a set of channels into composite values, that is, the weighted sums of separate channel readings. One of these weighted sums measures roughly the brightness of each pixel in the scene. The other composite values are linear combinations of the values of the separate channels, but some of the weights are negative and others positive. One of these other composite values represents the degree of greenness of the pixels and another might represent the degree of yellowness of vegetation or perhaps the wetness of the soil. Usually there are just three composite variables.

Something very much like the tasseled cap transformation could have arisen from strictly empirical observations. Principal component analysis could have provided the inspiration and guide for specifying the tasseled cap transformations. Principal component analysis (PCA) creates new variables as weighted sums of the different channel readings. Typically the first few components contain most of the information in the data so that four channels of LANDSAT MSS data or the six channels of the Thematic Mapper data may be reduced to just three principal components. The components higher than three are usually treated as being information less noise.

The weights used in principal component analysis are determined statistically from the data but it was soon observed that the first principal component typically corresponded to roughly equal weights. In other words, the data generally fall along the diagonal when channel values are plotted together. If the weights used in a weighted-sum transformation are equal then the values obtained are proportional to the sum of the channel values and hence correspond to "brightness."

Principal component analysis is equivalent to transforming the data to a new coordinate system with a new set of orthogonal axes. The tasseled cap transformation also corresponds to a transformation of the data to a new set of orthogonal axes. Although the tasseled cap transformation was inspired by the method of principal component analysis combined with generalization from empirical observations, the actual details had a more analytical basis.

The tasseled cap transformation was presented in 1976 by R. J. Kauth and G. S. Thomas of Environmental Research Institute of Michigan in an article titled "The Tasseled Cap—A Graphic Description of the Spectral-Temporal Development of Agricultural Crops as Seen by LANDSAT." This paper was published in the *Proceedings of the Symposium on Machine Processing of Remotely Sensed Data*, which was printed by Purdue University of West

Lafayette, Indiana. In this article Kauth and Thomas provide the rationale for the patterns found for LANDSAT data from crop lands as a function of the life cycle of the crop.

The data for bare soil can vary because of the character of the soil or as a function of sun angle in relation to the slope of the fields. In Figure 1.6 is the bare soil values A1 and B1 fall along a line through the origin. As a crop such as wheat emerges from dark soil there is an increase in reflectance in the near-IR band because of the reflectance of chlorophyll but a decrease in the red band because of chlorophyll's absorption of red light. Also the plants create shadows that result in lower readings from the soil. The shadowing of the soil by the plants will depend upon the orientation of the crop rows compared to the angle of the sun. If the sun is shining down the rows, the shadows on the soil will be less than if the crop rows are perpendicular to the direction of the sun.

The net result is that as the wheat plants grow the near-IR readings have a net gain and the red readings have a net loss as shown by the points A2 and A3 in Figure 1.6. For light soils the pattern may be different. The light soil has a high enough reflectance that as the wheat plants grows the reflectance of the plant even in the near-IR band is not enough replace the loss of reflectance from the light soil, so readings decrease in both bands but more so in the red band than in the near-IR band. This is shown by points B2 and B3 in the diagram.

When the wheat reaches maturity and in effect creates a closed canopy, the reflectance for the dark and light soil fields converge to points (i.e., A4 and B4). From this point the readings are the same for the dark and light soil fields. As the wheat crop ripens it starts to turn yellow, as in points A5 and B5. When the wheat is fully ripe the reflectance in the near-IR drops substantially as is the case for points (i.e., A6 and B6).

In a real wheat field there will be a distribution of the readings for bare soil along an ellipsoid as a result of variations in soil types and land angle relative to the sun angle. With growing wheat there will be variation among the wheat plants as to the stage in the life cycle. Thus the plot of the data for a wheat field may resemble a cap as shown in Figure 1.6, which is based on the depiction in Kauth and Thomas' article.

Kauth and Thomas (1976) define their tasseled cap transformation relying upon Figure 1.6. One component of their transformation is the weighted sum where the weights are statistically derived. In the original formulation the weights are not all equal. Later analyses simplified the transformation to be the sum of the channel readings and the result is characterized as "brightness." The second component is perpendicular to the first and its axis passes through the point of maturity of the plants, corresponding to points (i.e., A4 and B4) in Figure 1.6. The third component corresponds to an axis perpendicular to the first and second and passing through the point that represents ripened wheat, the "yellow stuff." The fourth component

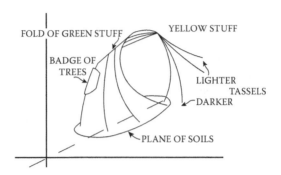

**FIGURE 1.6**
Tassel cap's common form (see Kauth and Thomas, 1976).

represents projection onto an axis perpendicular to the other three. Kauth and Thomas call it "non-such" and treat it as random noise.

The weights used by Kauth and Thomas for the tasseled cap transformation of LANDSAT MSS data are shown in the following table.

Weights for Tasseled Cap Transformation of LANDSAT MSS Data

| Component | Channel 1 | Channel 2 | Channel 3 | Channel 4 |
|-----------|-----------|-----------|-----------|-----------|
| Brightness | 0.433 | 0.632 | 0.586 | 0.264 |
| Greenness | −0.290 | −0.562 | 0.600 | 0.491 |
| Yellowness | −0.829 | 0.522 | −0.039 | 0.194 |
| "Non-such" | 0.223 | 0.012 | −0.543 | 0.810 |

The weights for the first component were found from data for Fayette County, Illinois. The green point was based upon a cluster for corn field data in Fayette County as well. Kauth and Thomas had no yellow point for the Fayette County data so they used the spectrum of yellow corn to approximate what the yellow point would be.

Crist and Cicone (1984) adapted the tasseled cap transformation to the six channels of Thematic Mapper data. The weights are different and the third component is taken to represent soil wetness rather than yellowness as in Kauth and Thomas' original formulation. The weights found by Crist and Cicone for the Thematic Mapper bands were:

Weights for Tasseled Cap Transformation of Thematic Mapper Data

| Component | Channel 1 | Channel 2 | Channel 3 | Channel 4 | Channel 5 | Channel 7 |
|-----------|-----------|-----------|-----------|-----------|-----------|-----------|
| Brightness | 0.3037 | 0.2793 | 0.4343 | 0.5585 | 0.5082 | 0.1863 |
| Greenness | −0.2848 | −0.2435 | −0.5436 | 0.7243 | 0.0840 | −0.1800 |
| Wetness | 0.1509 | 0.1793 | 0.3299 | 0.3406 | −0.7112 | −0.4572 |

## Geographic Information Systems (GIS)

GIS data represents real-world objects (roads, land use, elevation) with digital data. Real-world objects can be divided into two abstractions: discrete objects (a house) and continuous fields (rain fall amount or elevation). There are two broad methods used to store data in a GIS for both abstractions: raster and vector (see Figure 1.7).

Raster data type consists of rows and columns of cells where in each cell is stored a single value. Most often, raster data are images (raster images), but besides just color, the value recorded for each cell may be a discrete value (such as land use), a continuous value (such as rainfall), or a null value if no data is available. Although a raster cell stores a single value, it can be extended by using raster bands to represent RGB (red, green, blue) colors, color maps (a mapping between a thematic code and RGB value), or an extended attribute table with one row for each unique cell value. The resolution of the raster data set is its cell width in ground units. For example, in a lidar raster image, each cell may be a pixel that represents a 3 m × 3 m area. Usually cells represent square areas of the ground, but other shapes can also be used.

Vector data type uses geometries such as points (i.e., trees, water well, pole, plant, animal, other) and lines (series of point coordinates or polygons,

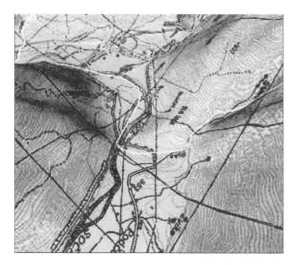

**FIGURE 1.7**
A representative of Digital Elevation Model (DEM) + Map (image) + Vector Data (layout layers).

also called areas [shapes bounded by lines]) to represent objects. Examples include property boundaries for a housing subdivision represented as polygons (lake, pond) and well locations represented as points. Vector features can be made to respect spatial integrity through the application of topology rules such as "polygons must not overlap." Vector data can also be used to represent continuously varying phenomena. Contour lines and triangulated irregular networks (TINs) are used to represent elevation or other continuously changing values. TINs record values at point locations, which are connected by lines to form an irregular mesh of triangles. The faces of the triangles represent the terrain surface.

There are advantages and disadvantages to using a raster or vector data model to represent reality. Raster data sets record a value for all points in the area covered, which may require more storage space than representing data in a vector format that can store data only where needed. Raster data also allows easy implementation of overlay operations, which are more difficult with vector data. Vector data can be displayed as vector graphics used on traditional maps, whereas raster data will appear as an image that may have a blocky appearance for object boundaries.

Additional nonspatial data can also be stored besides the spatial data represented by the coordinates of vector geometry or the position of a raster cell. In vector data, the additional data are attributes of the object. For example, a forest inventory polygon may also have an identifier value and information about tree species. In raster data the cell value can store attribute information, but it can also be used as an identifier that can relate to records in another table.

### Thematic Data Layers

ESRI, World Congress of Geography, USGS, NASA, and others use general features and data attributes that are tagged utilizing the international Feature and Attribute Coding Catalogue (FACC). These feature layers are cited as:

- Major road networks
- Railroad networks
- Hydrologic drainage systems
- Utility networks (cross-country pipelines and communication lines)
- Major airports
- Elevation contours
- Coastlines
- International boundaries
- Populated places
- Index of geographical name

## Geospatial Data Conversion

Remotely sensed data are based on raster format data structure (grid cell, x and y). Raster is a cellular-based data structure composed of rows and columns. The value of each cell represents the feature value, and groups of cells are used to represent each feature. The structure is commonly used to store image data (i.e., Landsat TM Data, digital elevation model [DEM]). However, vector data is a coordinate-based data structure commonly used to represent map features. Each linear feature is represented as a list of ordered x, y coordinates. Attributes are associated with each feature (as opposed to raster data structure, which associate attributes with a grid cell). Traditional vector data structures include double-digitized polygons and arc-node models (i.e., vegetation and soil map, river and stream, trees and wells). The raster and vector data are used for the purpose of spatial modeling and thematic mapping products.

Before doing any type of modeling, one needs to be familiar with the data. If the data is remotely sensed, the following steps will help in understanding the format, structure, and other characteristics associated with their data. This can be accomplished by using ERDAS-IMAGINE software (ERDAS: Earth Resource Data Analysis System); ENVI: Research Systems, Inc. (RSI); or EDRISI (Clark Labs, Clark University); ARCINFO (ESRI: Environmental Systems Research Institute, Inc.); ARCGIS (ARCMAP); or ARCVIEW software.

### Using ERDAS-IMAGINE Software

The software supports three different file formats:

- Thematic raster—These data types are qualitative and categorical representative data such as soil map, DEM, slope, etc.).
- Continues raster—These data are quantitative and have related continues values (multiple bands or single band, i.e., TM Data or SPOT, MODIS or radar).
- Vector data—These data have characteristics of encoding format composed of node and arcs (i.e., digital line graph (DLG) data, such as roads or hydrology, TIGER files).

If you are using a UNIX-based system (or environment), type *imagine* at the command line and open the viewer in ERDAS-IMAGINE. Within the viewer:

Click on File.
Click on Open.
Click on Raster.
Click filename to open (i.e., rmnp8000.img).

Rmnp8000.img is a portion of a Landsat TM image covering the eastern portion of the Rocky Mountain National Park, Colorado (Figure 1.8).

To view information about the TM imagery, within the viewer, click on Raster option and select band combinations (i.e., Band 4 [near IR—Red gun], Band 3 [red visible—green gun], Band 2 [green visible—blue gun]; see Figure 1.9). Using different band combinations can be very useful in describing various landscape features (i.e., forest types, river, agriculture field, etc.). From the viewer (Figure 1.9), click on Utility, then Layer Info to read information about the image (we can read only one band each time),

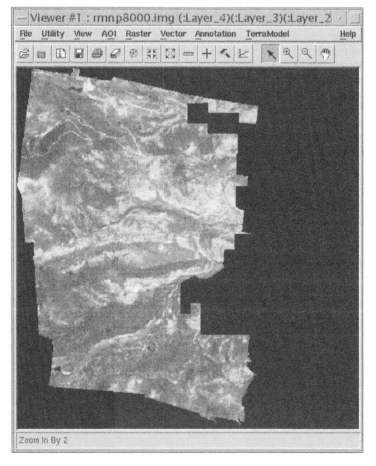

**FIGURE 1.8**
Landsat TM imagery of band 4, 3, and 2 for a study site, cover area of over 8000 ha, located in eastern part of the Rocky Mountain National Park, Colorado.

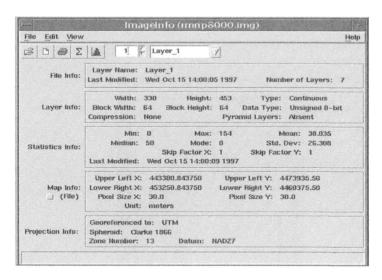

**FIGURE 1.9**
Image information used to describe Landsat TM imagery of the study site.

data structure, statistical information on band, MapInfo (size the area covered by image, geographic coordinates, pixel size, and other), and projection information (i.e., type of projection UTM and other information).

We also can convert remotely sensed imagery to ARCINFO-Grid data format for the purpose of extracting the digital number from each band of the Landsat TM, MODIS or SPOT, or radar. This can be done as follows: In ERDAS-IMAGINE use the Export option (Figure 1.10). This will convert the TM image file to seven layers (grids). For example, if we have an image called rmnp8000.img, the Import/Export option would produce seven GRID-ARCINFO files named: rmnp8000_L1, rmnp8000_L2, rmnp8000_L3, rmnp8000_L4, rmnp8000_L5, rmnp8000_L6, and rmnp8000_L7.

## Using ARCINFO Software

An alternative way to convert an image file format to a grid file type is to use the command IMAGEGRID in ARCINFO. To start ARCINFO type *arc* at the UNIX/MS Window prompt and then type *Arc: imagegrid rmnp8000.img rmnp8000.*

The imagegrid procedure converts an image of seven Landsat TM bands (rmnp8000.img) to a grid file format readable in GRID-ARCINFO. The output has seven layers, or grids named rmnp8000_L1, rmnp8000_L2, rmnp8000_L3, rmnp8000_L4, rmnp8000_L5, rmnp8000_L6, and rmnp8000_L7.

To convert a GRID-ARCINFO file format to ERDAS-IMAGNE file format, different steps are required. The following two examples create an

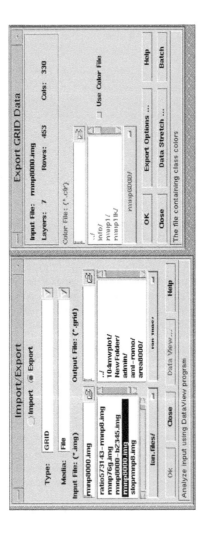

**FIGURE 1.10**
Using the Export option in ERDAS-IMAGINE.

image file called rmnp800.img from the seven grid files rmnp8000_L1, ..., rmnp800_L7.

1. Combine grid layers using the Makestack command in GRID-ARCINFO:

```
Grid: makestack rmnp8000 list rmnp8000_L1, rmnp8000_L2,
rmnp8000_L3, rmnp8000_L4, rmnp8000_L5, rmnp8000_L6, and
rmnp8000_L7
```

2. Convert Grid to image file using Gridimage command in ARCINFO:Arc: gridimage rmnp8000 # rmnp8000.img imagine

This step can be displayed and read as an image in ERDAS or ARCGIS, other.

### Select Area of Interest (Study Site)

After converting the Landsat TM imagery to a grid file format, one can produce a boundary file to represent a study site (area of interest). If the boundary file is in a vector format, in ARCINFO use the Clip command.

```
Arc: clip romo-veg rmnp8000-veg rmnp-site
```

The cover romo-veg is clipped using rmnp8000-veg and creates an output cover called rmnp-site.

If the file is in raster format (i.e., DEM), then in ARCINFO use the Latticeclip command:

```
Arc: latticeclip rmnp_dem rmnp8000-veg dem8000 or
Grid: gridclip rmnp-dem dem8000 cover rmnp8000-veg
```

In this example the grid rmnp-dem is clipped using rmnp8000-veg and creates the output grid dem8000.

### Topographic Data

Topographic data (elevation, slope, and aspect) are useful for spatial modeling in describing spatial variability within the landscape. These data can be derived from DEM using GRID-ARCINFO. If we have a DEM called dem8000, we can derive a grid of slopes and aspects using the commands:

```
Grid: slope8000 (dem8000, degree)
Grid: aspect8000 (dem8000)
```

Sometimes when dealing with vegetation studies, we may need to transform aspect data using the absolute value from due south (180°; high solar radiation and continues). The following example shows how to use absolute value as:

```
Grid: abs-aspect8000 = abs(180 - aspect8000)
```

Also, we may resample the raster data (remotely sensed data or DEM) to different cell sizes (pixel sizes). We can resample MODIS from 250 m² or 500 m² or Landsat TM data from 30 m × 30 m to SPOT data of 20 m × 20 m or 10 m × 10 m; or elevation data of 90 m × 90 m to 30 m × 30 m, as in this example:

```
Grid: new-dem30 = resample (dem90, 30)
```

## Global Positioning System (GPS)

This section focuses on another part of geospatial information sciences and technology, the global positioning system (GPS). It provides general information for the reader on what, how, where, and when we use GPS. (This information is cited from the following sources: (http://www.gps.gov/, http://www.navcen.uscg.gov/, http://www.fs.fed.us/database/gps/, and http://www8.garmin.com/aboutGPS/.) The GPS is a U.S. space-based radio navigation system that provides reliable positioning, navigation, and timing services to civilian users on a continuous worldwide basis—freely available to all. This is based on *U.S. Space-Based Positioning, Navigation, and Timing Policy* (December 2004). For anyone with a GPS receiver, the system will provide accurate location and time information in all weather, day and night, anywhere in the world.

The GPS is made up of three parts:

1. Satellites orbiting the Earth
2. Control and monitoring stations on Earth
3. The GPS receivers owned by users

GPS satellites broadcast signals from space that are picked up and identified by GPS receivers. Each GPS receiver then provides three-dimensional location (latitude, longitude, and altitude) plus the time. Individuals may purchase GPS handsets that are readily available through commercial retailers. Equipped with these GPS receivers, users can accurately locate where they are and easily navigate to where they want to go, whether walking, driving, flying, or boating. GPS has become a mainstay of transportation systems worldwide, providing navigation for aviation, ground, and maritime operations. Disaster relief and emergency services depend, upon GPS for location and timing capabilities in their life-saving missions. Everyday activities such as banking, mobile phone operations, and even the control of power grids, are facilitated by the accurate timing provided by GPS. Farmers, surveyors, geologists, and countless others perform their work more efficiently, safely, economically, and accurately using the free and open GPS signals.

The U.S. Air Force develops, maintains, and operates the space and control segments.

- The space segment consists of a nominal constellation of 24 operating satellites that transmit one-way signals that give the current GPS satellite position and time.
- The control segment consists of worldwide monitor and control stations that maintain the satellites in their proper orbits through occasional command maneuvers, and adjust the satellite clocks. It tracks the GPS satellites, uploads updated navigational data, and maintains the health and status of the satellite constellation.
- The user segment consists of the GPS receiver equipment, which receives the signals from the GPS satellites and uses the transmitted information to calculate the user's three-dimensional position and time.

### GPS Services

GPS satellites provide service to civilian and military users. The civilian service is freely available to all users on a continuous, worldwide basis. The military service is available to U.S. and Allied armed forces as well as approved government agencies. A variety of GPS augmentation systems and techniques are available to enhance system performance to meet specific user requirements. These improve signal availability, accuracy, and integrity, allowing even better performance than is possible using the basic GPS civilian service. The outstanding performance of GPS over many years has earned the confidence of millions of civil users worldwide. It has proven its dependability in the past and promises to be of benefit to users, throughout the world, far into the future.

### The GPS Satellite System and Facts

There are 24 satellites that make up the GPS space segment orbiting the Earth about 12,000 miles above us (see Figure 1.11). They are constantly moving, making two complete orbits in less than 24 hours. These satellites are travelling at speeds of roughly 7000 miles an hour. GPS satellites are powered by solar energy. They have backup batteries onboard to keep them running in the event of a solar eclipse, when there is no solar power. Small rocket boosters on each satellite keep them flying in the correct path.

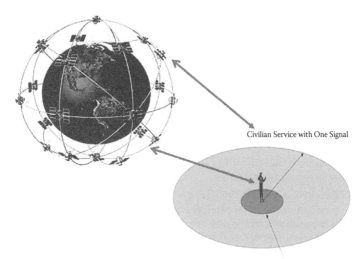

**FIGURE 1.11**
**(See color insert)** The GPS satellite system and stand-alone GPS Notional Horizontal Performance with New Signals. (Cited and modified from http://www8.garmin.com/about-GPS/ and http://www.gps.gov/systems/gps/.)

Facts about the GPS satellites (also called NAVSTAR, the official U.S. Department of Defense name for GPS):

- The first GPS satellite was launched in 1978.
- A full constellation of 24 satellites was achieved in 1994.
- Each satellite is built to last about 10 years. Replacements are constantly being built and launched into orbit.
- A GPS satellite weighs approximately 2000 pounds and is about 17 feet across with the solar panels extended.
- Transmitter power is only 50 watts or less.

## GPS Applications

In addition to using GPS for military, defense, intelligent, and homeland security activities, GPS also are used for outside military environments such as natural resources management, agriculture systems, environment and sustainability, water resources, urban planning, and research and teaching applications. GPS data collection systems provide decision makers with descriptive information and accurate positional data about items that are spread across many kilometers of terrain. By connecting position information with other types of data such as remotely sensed imagery, it is possible to analyze many environmental problems from a new perspective. Position data collected through GPS can be imported into GIS software, allowing spatial aspects to be analyzed with other information to create a far more complete understanding of a particular situation than might be possible through conventional means.

Aerial studies of some of the world's most impenetrable wilderness are conducted with the aid of GPS technology to evaluate an area's wildlife, terrain, and human infrastructure. By tagging imagery with GPS coordinates it is possible to evaluate conservation efforts and assist in strategy planning. Some countries collect and use mapping information to manage their regulatory programs such as the control of royalties from mining operations, delineation of borders, and the management of logging in their forests. GPS technology supports efforts to understand and forecast changes in the environment. By integrating GPS measurements into operational methods used by meteorologists, the atmosphere's water content can be determined, improving the accuracy of weather forecasts. The modernization of GPS will further enhance the support of GPS technology to the study and management of the world's environment. The United States is committed to implementing two additional civilian signals that will provide ecological and conservation applications with increased accuracy, availability, and reliability. Tropical rain forest ecology, for example, will benefit from the increased availability of GPS within heavy foliage areas and the reduction of spatial error in fine-scale vegetation mapping. Finally, the development and implementation of precision agriculture or site-specific farming has been made possible by combining GPS and GIS. These technologies enable the coupling of real-time data collection with accurate position information, leading to the efficient manipulation and analysis of large amounts of geospatial data. GPS-based applications in precision farming are being used for farm planning, field mapping, soil sampling, tractor guidance, crop scouting, variable rate applications, and yield mapping. GPS allows farmers to work during low visibility field conditions such as rain, dust, fog, and darkness. In conclusion, GPS has too many interesting applications to our planet Earth.

# References

American Society of Photogrammetry. 1983. *Manual of Remote Sensing*, 2nd ed., Vols. I and Vol. II (Editor-in-Chief Robert N. Colwell, Volume I Editor: David S. Simonett, Associate Editor: Fawwaz T. Ulaby; Volume II Editor: John E. Estes, Associate Editor: Gene A. Thorley). Falls Church, VA: American Society of Photogrammetry.

American Society of Photogrammetry. 1984. *Multilingual Dictionary of Remote Sensing and Photogrammetry, English Glossary and Dictionary French-German-Italian-Portuguese-Spanish-Russian*. Falls Church, VA: American Society of Photogrammetry.

Avery, T. E., and G. L. Berlin. 1992. *Fundamentals of Remote Sensing and Airphoto Interpretation*, 5th ed. New York: Macmillan.

Breslin, P, N. Frunzi, E. Napoleon, and T. Ormsby. 1998. *Getting to Know ArcView GIS, the Geographic Information System (GIS) for Everyone*. Redlands, CA: ESRI.

Carver, K. R. 1988. *SAR Synthetic Aperture RADAR–Earth Observing System*, Vol. IIf, NASA Instrument Panel Report, Washington D.C.

Cihlar, J., A. Belward, and Y. Govaerts. 1999. *Meteosat Second Generation Opportunities for Land Surface Research and Applications*. EUMETSAT Scientific Publications, EUM SP 01.

Crist, E. P. and R. C. Cicone. 1984a. Application of the tasseled cap concept to simulated Thematic Mapper data. *Photogrammetric Engineering and Remote Sensing*, pp. 343–352.

Crist, E. P., and R. C. Cicone. 1984b. *A Physically-Based Transformation of Thematic Mapper Data—The TM Tasseled Cap*, IEEE Transactions on Geoscience Remote Sensing, Vol. GE-22, 256–263.

Crist, E. P., and R. J. Kauth. 1986. The Tasseled Cap De-Mystified. *Photogrammetric Engineering and Remote Sensing* 52(1): 81–86.

Diner, D., G. P. Asner, R. Davies, Y. Knyazikhin, J. P. Muller, A. Nolin, B. Pinty, C. B. Schaaf, and J. Stroeve. 1999. New directions in Earth observing: Scientific applications of multi-angle remote sensing. *Bulletin of the American Meteorological Society* 80:2209–2228.

ERDAS. 1994. *ERDAS Field Guide*, 3rd ed. Atlanta: ERDAS.

ESRI. 1994. *Understanding GIS: The ARC/INFO Method*. Redlands, CA: Environmental System Research Institute.

Gobron, N., Pinty B., Mélin F., Taberner M. and Verstraete M. M. 2002. Sea Wide Field-of-View Sensor (SeaWiFS) - An Optimized FAPAR Algorithm - Theoretical Basis Document. Institute for Environment and Sustainability, EUR Report No. 20148 EN, p. 20.

Gobron, N., B. Pinty, M. M. Verstraete and Y. Govaerts. 1998. 'MERIS Level 2 Vegetation Index Algorithm Theoretical Basis Document', EC/JRC Publication EUR 18137 EN.

Gobron N., B. Pinty, and M. M. Verstraete. 1999. Spectral indices optimized for the characterization of land surfaces: Recent advances and preliminary results from VEGETATION, MERIS, and GLI. In *ALPS 99* (CNES, ed.), Meribel, France, January 18–22.

Gobron, N., B. Pinty, M. M. Verstraete, and Y. Govaerts. 1999. The MERIS Global Vegetation Index (MGVI): Description and Preliminary Application. *International Journal of Remote Sensing* 20:1917–1927.

Gobron, N., B. Pinty, M. M. Verstraete, and F. Mélin. 1999. *Development of a Vegetation Index Optimized for the SeaWiFS Instrument*. Algorithm Theoretical Basis Document, Joint Research Centre Publication EUR 18976 EN.

Govaerts, Y., M. M. Verstraete, B. Pinty, and N. Gobron. 1999. Designing optimal spectral indices: A feasibility and proof of concept study. *International Journal of Remote Sensing* 20:1853–1873.

Henderson, F. M., and A. J. Lewis, 1998. *Principles and Applications of Imaging Radar* (Manual of Remote Sensing, 3rd ed., Vol. 2). New York: John Wiley & Sons.

Jensen, J. R. 1996. *Introductory Digital Image Processing: A Remote Sensing Perspective,* 2nd ed. Upper Saddle River, NJ: Prentice Hall.

Jensen, J. R. 2000. *Remote Sensing of the Environment.* Upper Saddle River, NJ: Prentice Hall.

Kauth, R. J., and G. S. Thomas. 1976. The tasseled cap—A graphic description of the spectral-temporal development of agricultural crops as seen by LANDSAT. *Proceedings of the Symposium on Machine Processing of Remotely Sensed Data,* Purdue University of West Lafayette, Indiana, pp. 4B-41–4B-51.

Knorr W., N. Gobron, B. Pinty, M. M. Verstraete, and G. Dedieu. 1999. A new approach to derive monthly global fields of FAPAR from NOAA-AVHRR data. In *ALPS 99* (CNES, ed.), Meribel, France, January 18–22.

Leprieur, C., Verstraete, M. M., Pinty, B., Chehbouni, A. 1994. NOAA/AVHRR Vegetation Indices: Suitability for Monitoring Fractional Vegetation Cover of the Terrestrial Biosphere, in *Proc. of Physical Measurements and Signatures in Remote Sensing.*

Lillesand, T. M., and R. W. Kiefer. 1987. *Remote Sensing and Image Interpretation,* 2nd ed. New York: John Wiley & Sons.

McCloy, K. R. 2006. *Resource Management Information Systems: Remote Sensing, GIS and Modeling,* 2nd ed. New York: Taylor & Francis.

Mercer, B. 2004. DEMs Created from Airborne IFSAR—An update. *International Archives of Photogrammetry and Remote Sensing,* Vol. 35, part B. Istanbul, Turkey.

Moreira, J., M. Schwäbisch, C. Wimmer, M. Rombach, and J. Mura. 2001. Surface and ground topography determination in tropical rainforest areas using airborne interferometric SAR. *Photogrammetric Week 01,* D. Fritsch and R. Spiller, Eds. Heidelberg: Herbert Wichmann Verlag.

Ormsby, T., and J. Alvi. 1999. *Extending ArcView GIS: Teach Yourself to Use ArcView GIS Extension.* Redlands, CA: Environmental Systems Research Institute.

Pinty B., F. Roveda, M. M. Verstraete, N. Gobron, and Y. Govaerts. 2000. Estimating Surface Albedo from the Meteosat Data Archive: A Revisit. In *ALPS 99* (edited by CNES, ed.), Meribel, France, January 18–22.

Rencz, A. W., 1999. *Remote Sensing for the Earth Sciences* (Manual of Remote Sensing, 3rd ed., Vol. 3). New York: John Wiley & Sons.

Rosen, P. H. S. Joughin, I., Li, F., S. Madsen, E. Rodriguez, and R. Goldstein. 2000. Synthetic Aperture RADAR Interferometry. *Proceedings of the IEEE* 88(3).

Sabins, J. R., and F. Floyds. 1986. *Remote Sensing: Principles and Interpretation,* 2nd ed. New York: W. H. Freeman and Company.

Ustin, S., M. O. Smith, S. Jacquemoud, M. M. Verstraete, and Y. Govaerts. 1999. Geobotany: Vegetation mapping for Earth sciences. In *Remote Sensing for the Earth Sciences* (Manual of Remote Sensing, 3rd ed., Vol. 3), by Andrew N. Rencz, ed., 189–248. New York: John Wiley and Sons.

Verstraete, M. M. and B. Pinty. 1996. Designing optimal spectral indices for remote sensing applications, *IEEE Transactions on Geoscience and Remote Sensing,* 34, 1254–1265.

Verstraete, M. M., B. Pinty, and P. Curran. 1999. MERIS potential for land applications. *International Journal of Remote Sensing* 20:1747–1756.

Vogt, P., M. M. Verstraete, B. Pinty, M. Menenti, A. Caramagno, and M. Rast. 1999. *On the Retrieval Accuracy of the Albedo and BRF Fields: Potential of the LSPIM/PRISM Sensor*, Joint Research Centre Publication EUR 19016 EN.

Waring, R. H., J. B. Way, R. Hunt, L. Morrisey, K. J. Ranson, J. F. Weishampel, R. Oren, and S. E. Franklin. 1995. Biological toolbox–imaging RADAR for ecosystem studies. *BioScience* 45:715–723.

# 2

## Data Sampling Methods and Applications

This chapter covers data type and representation with examples.

### Data Representation

Data can be represented in a number of ways, including tables, charts, plots, and graphs. Data visualization of raw data can be very important in terms of data analysis, report writing, and presentation. All these steps can be called exploratory data analysis.

### Data Collection and Source of Errors

Data can be collected in different ways and from, for example, field data, existing data, and electronic, either remote (e.g., Landsat, satellite remotely sensed data) or close at hand (e.g., still video camera).
Errors associated with data collection can be related to many factors:

Instrument use and calibration

Data entry and registration

Rounding

Dislocation and labeling

Units of measure

Human error

Personal experience

#### Data Types

Ratio—A representative of quantities in terms of equal intervals and an absolute zero point of origin (0 to 1, 10, 100, 1000, etc.). Examples: age, height, vegetation or forest type, ethnic group, fuel loading and fire severity, weed invasion, wetland, gender (female vs. male).

Ratio input map data—Used with statistical operations such as diversity, minority, majority, sum, mean, median, minimum, maximum, and range.

Interval—A representative of quantities in terms of equal intervals or degrees of difference, but whose zero point is arbitrarily established. Examples: compass direction, time of the day, longitudes, latitudes, daily temperatures, rain.

Interval input map data—Can be used with statistical operations such as sum, mean, median, minimum, maximum, range, diversity, minority, and majority.

Nominal—A representative of qualities rather than quantities; discrete or categorical measurement types. Examples: soil type, fuel loading, forest type, road network type, water body (lake, stream, river), postal code, presence and absence.

Nominal input map data—Can be used with statistical operations such as diversity, minority, and majority.

Ordinal—A measurement represented by order or rank of value differences. Examples: smaller vs. larger, more vs. less, poor vs. moderate land use, better vs. worse, good vs. bad.

Ordinal input map data—Can be used with statistical operations such as sum, mean, median, minimum, maximum, range, diversity, minority, and majority.

---

## Sampling Methods and Applications

The method used in selecting a representative sample of ground reference data or remotely sensed data is considered an important part of any accuracy assessment (Card 1982). Thus, the selection of an appropriate sampling scheme is very important in estimating the structure of the error matrix. Poor choice of a sample design can result in inefficient estimates of the error matrix, which may or may not reflect the true accuracy of the remotely sensed data.

Husch et al. (1982, p. 150) stated that sampling design is determined by the kind of sampling units used, the number of sampling units employed, and the manner of selecting the sampling units and distributing them over the landscape area (forest, vegetation, rangeland, etc.) as well as the procedure for taking measurements and analyzing the results. In general, two sampling categories are commonly used:

1. Probability sampling—This type of sampling can be described as "the probability of selecting any sampling unit is assumed to be known prior to the actual sampling."

a. Simple random sampling (with and without replacement)
b. Stratified random sampling
c. Multistage sampling
d. Multiphase (double) sampling
e. Sampling with varying probabilities
f. Network sampling (rare characteristics: disease, people, patient records, etc.)
g. Cluster and systematic sampling
h. Purposive sampling (random selection of sample units without replacement)

2. Nonrandom Sampling—This sampling can be described as "the selection of sampling units are not governed by laws of chance (no probability, not subjective) but is done systematically or using personal judgment."

a. Selective sampling—Personal selection of sample
b. Systematic sampling—Sampling units are spaced at fixed intervals throughout the population
c. Purposive (or adaptive) sampling—Sample with a purpose in mind

## Sampling Designs

The development of the sampling and plot designs is complicated by the diversity of variables and indicators to be assessed, the need to assess the landscape scale ecosystem resources at a range of scales, the need to monitor the indicators over time, and the need to do so efficiently. Sampling techniques commonly used to assess the accuracy of a classification procedure includes simple random sampling, systematic sampling, stratified random sampling, and cluster sampling (Congalton 1991; Kalkhan 1994; Kalkhan et al. 1997). Multiphase (i.e., double) sampling is also used to assess thematic mapping accuracy (Kalkhan et al. 1998). The following provide simplistic descriptions of the most common sampling designs used by researchers, resource management, and so forth into the integration of geospatial information for natural, landscape-scale, and environmental studies.

### Simple Random Sampling

Simple random sampling (SRS) is the fundamental selection method (Husch et al. 1982). Husch et al. (1982, p. 162) point out that all other sampling procedures are modifications of simple random sampling, which are designed to achieve greater economy or precision. A simple random sample is one in which the $n$ sample points are chosen independently and uniformly within the region "1 or 2 or 3." Simple random sampling has the advantage over other designs in that it is easy to apply and provides satis-factory results in evaluating the accuracy assessment of remotely sensed data. SRS is the most basic design and should work well in describing the spatial continuity of almost any variable and allows one to make infer-ences about the population of interest. There some disadvantage when using SRS. First, the sample size within each thematic class is proportional to its area (Congalton 1991). Second, there is increased cost associated with traveling between sample points. Also some areas within the landscape (forest landscape such as rare habitat, aspen stand) may not be sampled. For example, if you are sampling a population with several vegetation types, the number of samples in each vegetation type will be proportional to their area.

The steps to generate SRS is to develop a list of groups of people, trees, and plants by name or number, then randomly selecting a representative sample from the overall population.

## Stratified Random Sampling

Stratified random sampling (STRS) has the advantage over simple random sampling and systematic sampling in that a small number of samples are selected from each category to ensure reliable estimates. Stratified random sampling can be easy to implement by dividing the group (i.e., student, veg-etation or soil types, age of trees, height classes), then randomly selecting from each unit or stratum (Figure 2.1).

For example, in a stratified random sample, the population of interest is subdivided into $K$ homogeneous subgroups, or strata, and a sample of size, $n_i$, is selected from each stratum such that $n = \sum_{i=1}^{K} n_i$. Within strata, sample points can be chosen randomly or systematically. The biggest advantage of stratified random sampling is it ensures that all parts of a population are sampled equally well.

## Systematic Sampling

In contrast, systematic sampling has the advantage of spreading the sam-pling units uniformly throughout the study area, thereby reducing the cost of the survey by minimizing travel time. Also, by using this design, a single primary unit consists of secondary units spaced in some system-atic fashion throughout the population (Thompson 1992). Although this design minimizes travel time, it has other problems associated with it, especially from a geostatistical spatial modeling point of view. As men-tioned previously, the minimum spacing between sample points, deter-mines the resolution of the spatial model. If the spacing of the sample

Stratified with Simple Random Sampling

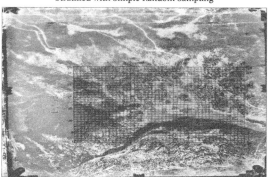

**FIGURE 2.1**
**(See color insert)** Example using air photo for stratify vegetation cover type and within each cover, a set of random points allocated.

plots is greater than the scale of the pattern of interest, it may not be possible to spatially model the variable of interest. Thus, it is important that the spacing between sample points be selected in such a way that it will capture the spatial variability associated with the variable being modeled. Systematic sampling has a similar disadvantage as simple random sampling in that the number of samples in a given vegetation type will be proportional to its area. Here are the steps we need to follow to achieve a systematic random sample:

1. Number the units in the population from 1 to N (e.g., 1 to 100)
2. Decide on the $n$ (sample size) that you want or need (e.g., 17)
3. $k = N/n$ = Interval Size (i.e., 6)
4. Randomly select an integer between 1 and $k$ (e.g., from 1 to 6 chose number 3)
5. Then take every $k$th unit (start from number 3 and take every 6th unit)

The following table is a procedure describing the systematic random sampling.

| 1 | 2 | 3 | 4 | 5 | 6 | 7 | 8 | X | 10 | 11 | 12 | 13 | 14 | X | 16 | 17 | 18 | 19 | 20 |
|---|---|---|---|---|---|---|---|---|----|----|----|----|----|---|----|----|----|----|-----|
| X | 22 | 23 | 24 | 25 | 26 | X | 28 | 29 | 30 | 31 | 32 | X | 34 | 35 | 36 | 37 | 38 | X | 40 |
| 41 | 42 | 43 | 44 | X | 46 | 47 | 48 | 49 | 50 | X | 52 | 53 | 54 | 55 | 56 | X | 58 | 59 | 60 |
| 61 | 62 | X | 64 | 65 | 66 | 67 | 68 | X | 70 | 71 | 72 | 73 | 74 | X | 76 | 77 | 78 | 79 | 80 |
| X | 82 | 83 | 84 | 85 | 86 | X | 88 | 89 | 90 | 91 | 92 | X | 94 | 95 | 96 | 97 | 98 | X | 100 |

### Nonaligned Systematic Sample

A nonaligned systematic sample is one in which the region of interest, for example, vegetation landscape, is divided into $m$-square, or rectangular strata, and a uniform random sample of size $k$ are chosen, so $n = km$. This design basically combines simple random sampling and systematic sampling, in that the sample points are spread uniformly and independently throughout the region A. Since there is no minimum distance between sample points, this design should capture the spatial variability in most populations.

### Cluster Sampling

Cluster sampling has also been used to assess the accuracy of remotely sensed data (Congalton 1984, 1991; Walsh and Burk 1993; Czaplewski 1992). Cluster sampling is generally recommended when travel time is an important component of the overall cost of the survey (Husch et al. 1982). Instead of collecting information on only one pixel of remote sensing image at a given location, information is collected on several pixels, thereby maximizing the information collected per unit travel time. Each remotely sensed pixel represents a cluster and could be a land cover type, such as forest or crops. As stated by Thompson (1992, p. 113): "The key point in any of the systematic or clustered arrangements is that whenever any secondary unit of a primary unit is included in the sample, all the secondary units of that primary unit are included." The general outline for cluster sampling is:

1. Divide population into clusters (usually along geographic boundaries, school within city)
2. Randomly sample clusters (classes)
3. Measure all units within sampled clusters (students)

### Multiphase (Double) Sampling

One method to reduce the cost of collecting data is to use aerial photography. Aerial photographs are less expensive than ground data, but their degree of misclassification tends to be higher than field data (Czaplewski 1992; Kalkhan 1994; Kalkhan et al. 1998). As an alternative, some combination of aerial photography and ground data—double sampling—could be used to provide more efficient estimates of the accuracy of remotely sensed data at a minimal cost (Czaplewski 1992; Kalkhan et al. 1995, 1998).

Neyman (1938) was the first to propose double sampling for estimating stratum weight in large regional inventories. Since Neyman's work in 1938, double sampling has been used in a variety of natural resources applications. In most of these applications, the variable of interest is costly to obtain; whereas an auxiliary variable, which is highly correlated to the variable of interest, is relatively inexpensive to collect. Regression or ratio estimators are then used to obtain estimates of the desired population parameters (Tryfos 1996).

---

## Double Sampling and Mapping Accuracy

The aforementioned techniques also can be applied to assess the accuracy of remotely sensed data. In the first phase, a large number of sample points are randomly located on aerial photographs. At each sample point, the cover type is recorded. The sample points are then georeferenced to the Landsat Thematic Mapper (TM) classification map and the cover types associated with each of the selected pixels are reported. This information is used to construct an error matrix describing the accuracy of TM classification imagery with respect to the aerial photography. In the second phase, a random subsample is selected from the first phase sample points and located in the field to verify the accuracy of the aerial photographs and the remotely sensed imagery. The information collected in the second phase is used to correct the misclassification error associated with the sample in the first phase using a composite estimator (Czaplewski 1992; Kalkhan et al. 1995, 1998; Figure 2.2).

The composite estimator is defined as a procedure that combines two estimates (i.e., two phases of sampling) into a single, more efficient estimate (Maybeck 1979, p. 217). In the univariate case, the composite estimator is the weighted sum of two estimates, with weights inversely proportional to the estimated variance of each prior estimate (Gregoire and Walters 1987). Furthermore, the composite estimator will have a variance somewhere in between the largest and smallest variance associated with the individual components making up the estimator (Burk et al. 1981). Composite estimators have played an important role in forest inventories and continue to generate interest because of the widely accepted belief that the more relevant information used by an estimator, the better the resulting estimate will be (Gregoire and Walters 1987). The use of composite estimators can result in significant savings in the amount of field data required (Green and Strawderman 1986).

Czaplewski (1992) proposed the composite estimator provided by Maybeck (1979 pp. 217, 247) as a statistical method to improve the precision of the estimates of land-cover map classification. The composite estimator provides considerable flexibility to accommodate complex sampling designs for cross-classified census and sample survey data (Czaplewski 1997). The addition of aerial photography as an intermediate phase creates the double sampling, which can be used in conjunction with the composite estimators to obtain an unbiased estimate, and reduce the amount of fieldwork, sample sizes, and the time and cost of assessing the accuracy of thematic maps.

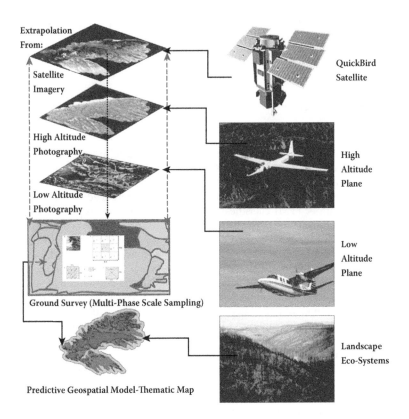

**FIGURE 2.2**
**(See color insert)** Multiphase, double sampling design using Landsat remote sensing, high spatial sensor, aerial photo, and ground survey for modeling and mapping of the landscape-scale assessment. (Adopted and modified from Kalkhan, M. A., et al., 1998, *International Journal of Remote Sensing* 19:2049–2060; Kalkhan et al. 2001, Proceeding of the Eight Forest Remote Sensing Application Conference (RS 2000), April 10-14, 2000, Albuquerque, New Mexico, 11 pp). Multiscale sampling and pixel nested plot designs for landscape-scale assessment. (Adopted and modified from Kalkhan et al. 2001; Kalkhan, M. A., et al., 2007, *Diversity and Distribution* 13:379–388; Kalkhan, M. A., et al., 2007, *Journal of Applied Soil Ecology* 37:622–634.)

## Pixel Nested Plot (PNP): Case Study

Geospatial statistical modeling and thematic maps have recently emerged as effective tools for the management of natural areas at the landscape scale. Traditional methods for the collection of field data pertaining to questions of landscape were developed without consideration for the parameters of these applications. An alternative field sampling design based on smaller unbiased random plot and subplot locations is called the pixel nested plot (PNP). We demonstrated the applicability of the PNP design of 15 m × 15 m to assess patterns of plant diversity and species richness across the landscape at Rocky Mountain National Park (RMNP), Colorado, in a time- and cost-efficient manner for field data collection (Figure 2.3 and Figure 2.4). Our results produced comparable results to the previous Beaver Meadow study (BMS) area within RMNP, where there was a demonstrated focus of plant diversity. Our study used the smaller PNP sampling design for field data collection, which could be linked to geospatial information data and could be used for landscape-scale analyses and assessment applications. In 2003, we established 61 PNPs in the eastern region of RMNP. We present a comparison between this approach using a subsample of 19 PNPs from this data set and 20 modified Whittaker nested plots (MWNPs) of 20 m × 50 m that were collected in the BMS area. The PNP captured 266 unique plant species, whereas the MWNP captured 275 unique species. Based on a comparison of PNP and MWNP in the Beaver Meadows area, the PNP required less time and area sampled to achieve a similar number of species sampled. Using the PNP approach for data collection can facilitate the ecological monitoring of these vulnerable areas at the landscape scale in a time- efficient and therefore cost-efficient manner (Figure 2.3).

This satisfies a need to measure patterns of biodiversity and other ecological variables at finer scales to create a direct link to the geospatial information, which can then be used to extrapolate across the landscape from fine to coarse scales. The PNP can be used to accommodate this need for

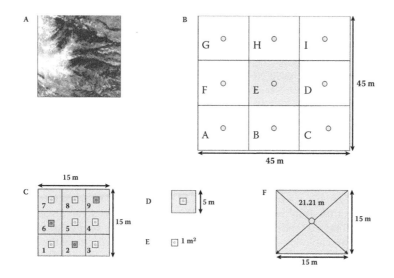

**FIGURE 2.3**
**(See color insert)** Pixel nested plots (PNP) were used to collect vegetation, biotic, and abiotic data. A, B, C and E represent size of 15m x 15m, D represents size of 5m x 5m, and F represents 1m x 1m. (For more information see Kalkhan et al., 2007a, 2007b.)

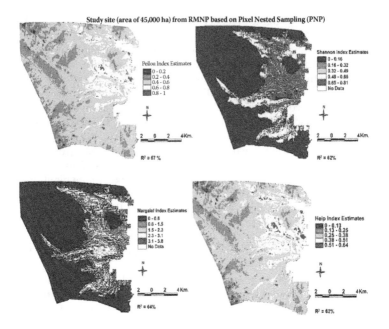

**FIGURE 2.4**
**(See color insert)** Geospatial model using GLM and regression classification trees to model the residual to predict plant diversity indices for an area cover about 45,000 ha with elevation range from 7000 ft to 10,000 ft. within the eastern region of the Rocky Mountain National Park (RMNP), Colorado. About 121 pixel nested plots (PNPs) were established by Colorado State University research work at RMNP during the summer field seasons 2003 and 2004. (Based in part on Kalkhan, 2006. ASPRS 2006 Annual Conference, *Prospecting for Geospatial Information Integration*, Reno, Nevada, May 1-5, 2006 [Contributed Papers: Image Classification–Spectral-Spatial Techniques].)

ecological–environmental forecasting, and for the inventory and monitoring of patterns of plant diversity (Figure 2.4) and invasions of exotic plant species (Kalkhan et al., 2007b).

PNP design is a nested plot type, an equilateral quadrangle, always orientated to the north and can be applied at numerous different spatial resolutions or sizes (i.e., 15 m × 15 m, 20 m × 20 m, 30 m × 30 m, 60 m × 60 m, or larger). In this study we use a 15 m × 15 m size because of the diverse and complex forested landscapes found in the RMNP and as a rapid assessment for biotic and abiotic parameters. The multiscale nested design of the PNP is also complementary in terms of the ground spatial resolution to other satellite sensors such as IKONOS, Orbview 3, QuickBird, RADARSAT, Landsat-7 ETM+ (panachromatic band with 15 m²), and SPOT-5, which have even finer scale resolutions (ranging from 1 m² to 20 m²). Each of the nested subplots is an equilateral quadrangle, all the way down to the 1 m² level of ground resolution. This design can be used easily and enables researchers and resource managers to create a direct link between data collected in the field with GPS locations and remotely sensed GIS data at different spatial resolutions (Kalkhan et al. 2007a). Since the link to geospatial information data for the PNP design is strong, we believe that our plot design improves on other square plot designs by being able to estimate patterns of biological diversity and species richness using fine-scale resolution down to the 1 m² level.

---

## Plot Design

Plot shape is commonly represented by a geometrical shape (square, rectangular, circular) or Point or transect (plotless). Square plots are preferred for the purpose of geospatial modeling and mapping since the analysts use remotely sensed data and GIS data, which are represented by pixels (grid cell: x, y coordinate), unless conditions specifically require otherwise (e.g., narrow plots for riparian strips or soft herbaceous vegetation that is easily damaged by trampling).

### Issues

Factors to consider are:

- Whether trampling is an issue (use narrow rectangles rather than squares or circles)
- Ease of setting up (e.g., circular plots are difficult to establish in tall vegetation)
- Cost effectiveness (ease of recording, number of plots relative to the degree of precision needed)

### Characteristics of Different Plot Shapes

Different plot shapes are presented in Figure 2.5. Square plots are the most commonly used and are relatively easy to set up. They have low edge effects relative to area of the plot when compared with rectangular plots. However, they can also sustain significant trampling of plants, which may affect some attributes being measured such as cover, compared with narrow rectangular plots.

Circular plots have similar characteristics to square ones. They have advantages with certain kinds of measurements such as basal area, which are determined by sittings from a fixed point (these BA samples are actually from irregular circular plots, not from plots for which the perimeter of the

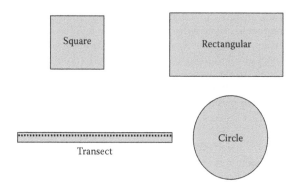

**FIGURE 2.5**
Different plot shapes layout.

circle is marked out on the ground). It can be difficult to mark the boundaries where the vegetation is tall.

Rectangular plots, including "transect" plots, allow greater access to the plot with relatively less trampling effects compared with square or circular plots. However, they have a relatively greater perimeter to area, which increases the risk of edge effects, such as the decision of whether a plant is in or out of the plot. Rectangular plots have advantages in sampling vegetation features that are by nature long and narrow in shape, for example, riparian habitats. When using rectangular or transect plots, the orientation of the plot relative to the surrounding environment may become a factor. There are several considerations. Which you choose depends on the purpose of the study or the kinds of analyses intended for the data.

- Orient the long axis of the plot *across* any environmental gradient (for example, slope) that is part of the homogeneous unit being sampled. This will assist by increasing the within-sample variation, and making later comparisons with other samples of the type easier. If the number of samples is determined by statistical or other quantitative analysis, then this approach will tend to reduce the number of samples needed compared to the next alternative. It tends to produce a conservative result with respect to recognizing vegetation types, that is, it tends to aggregate results rather than subdivide them.

- Orient the long axis parallel to within-type environmental gradients (such as slope), thus minimizing the within-sample variation. This process narrows the variation included in the sample, but tends to increase the apparent distinctness of samples from what are thought to be a single type thus tending to increase the number of types recognized in subsequent analyses.

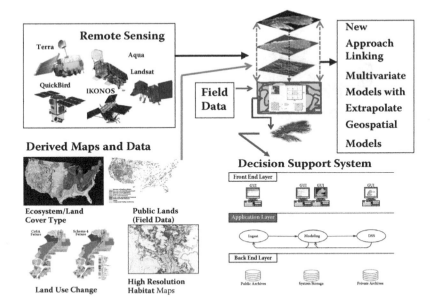

**FIGURE 1.1**
Multiple steps that can be used to integrate geospatial information sciences, geospatial forecasting systems, and mapping with decision support system for natural resources management, ecological forecasting, and environmental sustainability applications.

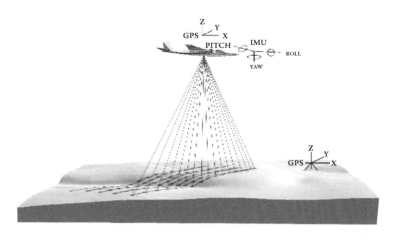

**FIGURE 1.5**
Lidar sensor operation. (This figure is property of and cited from http://forsys.cfr.washington.edu/JFSP06/lidar_technology.htm.)

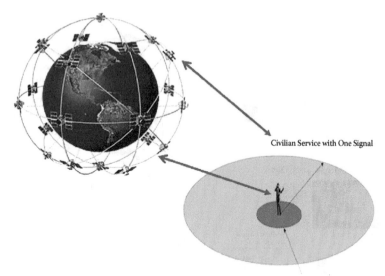

Civilian Service with One Signal

Civilian Service with New Signals

**FIGURE 1.11**
The GPS satellite system and stand-alone GPS Notional Horizontal Performance with New Signals. (Cited and modified from http://www8.garmin.com/aboutGPS/ and http://www.gps.gov/systems/gps/.)

## Stratified with Simple Random Sampling

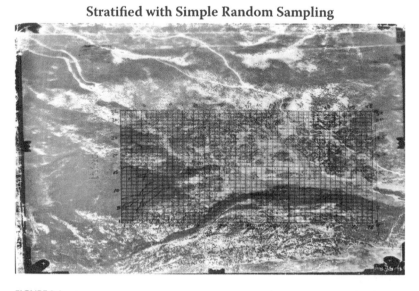

**FIGURE 2.1**
Example using air photo for stratify vegetation cover type and within each cover, a set of random points allocated.

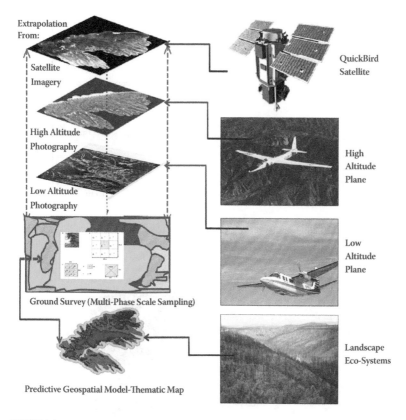

**FIGURE 2.2**

Multiphase, double sampling design using Landsat remote sensing, high spatial sensor, aerial photo, and ground survey for modeling and mapping of the landscape-scale assessment. (Adopted and modified from Kalkhan, M. A., et al., 1998, *International Journal of Remote Sensing* 19:2049–2060; Kalkhan et al. 2001, Proceeding of the Eight Forest Remote Sensing Application Conference (RS 2000), April 10-14, 2000, Albuquerque, New Mexico, 11 pp). Multiscale sampling and pixel nested plot designs for landscape-scale assessment. (Adopted and modified from Kalkhan et al. 2001; Kalkhan, M. A., et al., 2007, *Diversity and Distribution* 13:379–388; Kalkhan, M. A., et al., 2007, *Journal of Applied Soil Ecology* 37:622–634.)

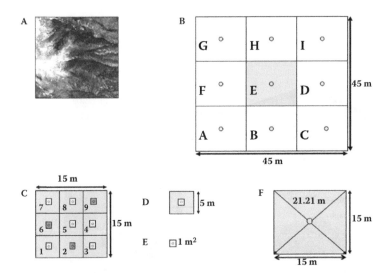

**FIGURE 2.3**
Pixel nested plots (PNP) were used to collect vegetation, biotic, and abiotic data. A, B, C and E represent size of 15m x 15m, D represents size of 5m x 5m, and F represents 1m x 1m. (For more information see Kalkhan et al., 2007a, 2007b.)

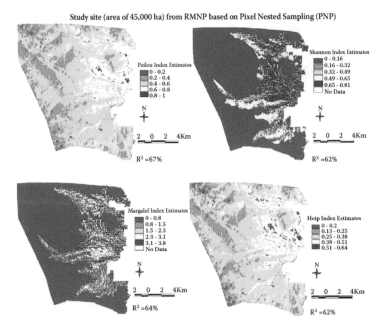

**FIGURE 2.4**
Geospatial model using GLM and regression classification trees to model the residual to predict plant diversity indices for an area cover about 45,000 ha with elevation range from 7000 ft to 10,000 ft. within the eastern region of the Rocky Mountain National Park (RMNP), Colorado. About 121 pixel nested plots (PNPs) were established by Colorado State University research work at RMNP during the summer field seasons 2003 and 2004. (Based in part on Kalkhan, 2006. ASPRS 2006 Annual Conference, *Prospecting for Geospatial Information Integration*, Reno, Nevada, May 1-5, 2006 [Contributed Papers: Image Classification–Spectral-Spatial Techniques].)

- Orient the long axis at random relative to within-type environmental gradients. By taking several samples within the type, the variance due to the included gradient will be incorporated in the aggregated sample of several transects for the location. This approach is similar to the first one.

## Plot Size

Plot size refers to the area (m²) covered by the sample plot. Plot size should vary with the physical dimensions of the things being sampled. As a guide, vegetation or forest size can vary as:

Smaller sizes <1 m high use plots from 1 m² (1 m × 1 m) to 4 cm² (2 cm × 2 cm; or 0.001256 m² [0.02 m × 0.02 m]).

Small size >1 (1, or 2, or 3, 5) m²

Medium size >10 (10, 15, 20) m²

Large size (30, 60) m²

Larger than 60 (90, 120, 240, 500, 1000) m²

### *What to Record*

Record the dimensions of the plot and the type of plot. If circular, give the radius in meters. If rectangular (including square) give the length and width. If point-centered or "plotless," indicate by naming method, and if transect, give the length.

### *Issues*

Plot size should remain as constant as possible for each vegetation type being sampled. Changing the area of a plot may affect some of the statistics for the survey, for example, within or between sample variances of certain statistics. However, it is appropriate to change the shape (as long as size [area] is not changed) of a plot so that its boundaries do not cross into adjoining vegetation types. For example, use narrow rectangular plots when sampling vegetation that grows in long narrow strips such as riparian vegetation.

Different sizes of plants/trees are usually better sampled by different sizes of plot. That is, large plants like trees require large plots, whereas shrubs require smaller plots, and herbs and mosses require smaller ones still; that is, nested plots may need to be part of your sampling design (e.g., when sampling stratified vegetation, the overstory may require a larger plot size than the midstratum, than the ground stratum). When doing detailed studies in previously unstudied vegetation types, it may be necessary to determine optimal plot size based on sampling sets of plots for the attribute(s) in

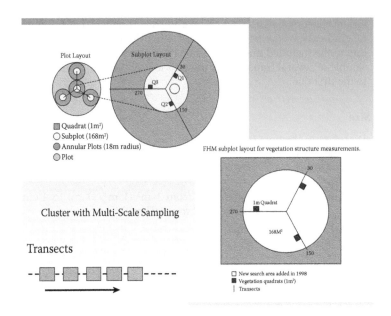

**FIGURE 2.6**
Different plot designs and shapes for forest and vegetation survey.

question and plotting the cumulative means of these sets against the cumulative area. The process is repeated until the fluctuation in the mean value sampled ceases or is reduced to negligible.

If quantitative floristic analyses are planned, then species accumulation curves should be determined and optimal plot size deduced from them. To determine the mean sizes of crowns and crown gaps, several transect 50 m or more long are recommended, rather than the use of extra-large plots. In sites dominated by the ground layer (grasslands, low shrublands, mosslands, etc.) several 1 m$^2$–0.01m$^2$ to total area of perhaps 50 m$^2$–0.5 m$^2$ or transects 1–20 m long should be used to collect foliage cover and plant height data (Figure 2.6).

## References

Breslin, P, N. Frunzi, E. Napoleon, and T. Ormsby. 1998. *Getting to Know Arc View GIS, the Geographic Information System (GIS) for Everyone*. Redlands, CA: ESRI.
Brown, J. K. 1974. A planar intersects method for sampling fuel volume and surface are. *Forest Science* 17:96–102.

Burk, R. E., Hansen, M. H., and Ek, A. R. 1981. Combing sources of information for improved in-place inventory statistics, In In-place resources inventories: principles and practices. Proceeding of National Workshop, edited by T. B. Brann, L. O. House IV, and H. G. Lund (Washington, DC: Society American Foresters), Publication No. 82±02, pp. 413±420.

Card, D. H. 1982. Using known map categorical marginal frequencies to improve map accuracy. *Photogrammetric Engineering and Remote Sensing* 48:431–439.

Cochran, W. G. 1977. *Sampling Techniques*, 3rd ed. New York: John Wiley & Sons.

Congalton, R. G. 1984. *A Comparison of Five Sampling Schemes Used in Assessing the Accuracy of Land Cover/Land Use Maps Derived from Remotely Sensed Data*. Ph.D. dissertation, Virginia Polytechnic Institute and State University. Blacksburg, VA.

Congalton, R. G. 1991. A review of assessing the accuracy of classifications of remotely sensed data. *Remote Sensing of Environment* 37:35–46.

Czaplewski, R. L. 1992. Accuracy assessment of remotely sensed classification with multi-phase sampling and the multivariate composite estimator. *Proceedings of the 16th International Biometrics Conference*, Hamilton, New Zealand, December 7–11, 2:22.

Czaplewski, R. L. 1997. Assessment of classification accuracy and extent estimates for land cover map with double sampling. In *Spatial Accuracy in Natural Resource Analysis*, H. Todd Mowrer and R. Congalton, eds. Chelsea, MI: Ann Arbor Press.

ESRI. 1994. *Understanding GIS: The ARC/INFO Method* Redlands, CA: Environmental System Research Institute.

Freese, F. 1962. *Elementary Forest Sampling*. Agriculture Handbook No. 232, USDA Forest Service.

Green, E. J., and W. E. Strawderman. 1986. Reducing sample size through the use of a composite estimators: An application to timber volume estimation. *Canadian Journal of Forest Resources* 16:1116–1118.

Gregoire, T. M., and D. K. Walters. 1987. Composite vector estimators derived by weighting inversely proportional to variance. *Canadian Journal of Forest Research* 18:282–284.

Greig-Smith, P. 1983. *Quantitative Plant Ecology* (Studies in Ecology, Vol. 9). Oxford: Blackwell Scientific.

Husch, B., C. I. Miller, and T. W. Beers. 1982. *Forest Mensuration*, 2nd ed. New York: John Wiley & Sons.

Kalkhan, M. A. 1994. Statistical properties of six accuracy indices using simple random and stratified random sampling: An application in remote sensing. Ph.D. dissertation, Colorado State University.

Kalkhan, M. 2006. Pixel Nested Plot: A Landscape-Scale Assessment Approach. ASPRS 2006 Annual Conference, "Prospecting for Geospatial Information Integration," Reno, Nevada, May 1-5, 2006 (Contributed Papers: Image Classification–Spectral-Spatial Techniques).

Kalkhan, M. A., R. M. Reich, and R. L. Czaplweski. 1996. Statistical properties of measures of association and the Kappa statistic for assessing the accuracy of remotely sensed data using double sampling. In *Spatial Accuracy Assessment in Natural Resources and Environmental Sciences: Second International Symposium*, May 21–23, Fort Collins, Colorado, USA, pp. 467–476.

Kalkhan, M. A., R. M. Reich, and R. L. Czaplewski. 1997. Variance estimates and confidence intervals for the Kappa measure of classification accuracy. *Canadian Journal of Remote Sensing* 9:210–216.

Kalkhan, M. A., R. M. Reich, and T. J. Stohlgren. 1998. Assessing the accuracy of Landsat Thematic Mapper map using double sampling. *International Journal of Remote Sensing* 19:2049–2060.

Kalkhan, M. A., E. J. Stafford, and T. J. Stohlgren. 2007b. Rapid plant diversity assessment using a pixel nested plot design: A case study in Beaver Meadows, Rocky Mountain National Park, Colorado, USA. *Diversity and Distribution* 13:379–388.

Kalkhan, M. A., E. J. Stafford, P. J. Woodley, and T. J. Stohlgren. 2007a. Exotic plant species invasion and associated abiotic variables in Rocky Mountain National Park, Colorado, USA. *Journal of Applied Soil Ecology* 37:622–634.

Kalkhan, M. A., T. J. Stohlgren, and M. Coughneour. 1995. An investigation of bio-diversity and Landscape-scale gape patterns using double sampling: A GIS approach. *In Ninth International Symposium on Geographic Information Systems for Natural Resources, Environment and Land Information Management*, March 27–30, Vancouver, British Columbia, Canada, pp. 708–712.

Kalkhan, M. A., T. J. Stohlgren, G. W. Chong, L. D. Schell, and R. M. Reich. 2001. A predictive spatial model of plant diversity: Integration of Remotely Sensed data, GIS, and Spatial statistics. Proceeding of the Eight Forest Remote Sensing Application Conference (RS 2000), April 10-14, 2000, Albuquerque, New Mexico, 11 pp. CD-ROMs Publications (ISBN 1-57083-062-2).

Kent, M., and P. Coker. 1992. *Vegetation Description and Analysis: A Practical Approach.* Boca Raton, FL: CRC Press.

Kershaw, K. A. 1966. *Quantitative and Dynamic Ecology.* London: Edward Arnold.

Laurini, R., and D. Thompson. 1992. Fundamentals of spatial information systems. New York: Academic Press.

Maybeck, P. S. 1979. *Stochastic Models, Estimation, and Control*, Vol. 1. New York: Academic Press.

Mueller-Dombois, D., and H. Ellenberg. 1974. *Aims and Methods of Vegetation Ecology.* New York: John Wiley & Sons.

Neyman, J., 1938. Contribution to the theory of sampling human population. *Journal of American Statistic Association* 33:101–116. (As cited in Cochran, W. G. 1977, *Sampling Techniques*, 3rd ed., New York, John Wiley & Sons.)

Ormsby, T., and J. Alvi. 1999. *Extending ArcView GIS: Teach Yourself to Use ArcView GIS Extension.* Redlands, CA: Environmental Systems Research Institute.

Reich, R. M., C. Aguirre-Bravo, and M. Iqbal. 1992. Optimal plot size for sampling coniferous forests in El Salto, Durango, Mexico. *Agrociencia* 2:93–106.

Reich, R. M., and L. G. Arvanitis. 1992. Sampling unit, spatial distribution of trees, and precision. *Northern Journal of Applied Forestry* 9:3–6.

Scheaffer, R. L., W. Mendenhall III, and L. Ott. 1996. *Elementary Survey Sampling.* Belmont, CA: Duxbury Press.

Schloeder, C. A., N. E. Zimmermann, and M. J. Jacobs. 2001. Comparison of methods for interpolating soil properties using limited data. *American Society of Soil Science Journal* 65:470–479.

Schreuder, H. T., T. G. Gregoire, and G. B. Wood. 1993. *Sampling methods for multiresource forest inventory.* New York: John Wiley & Sons.

Schreuder, H. T., M. S. Williams, C. Aguirre-Bravo, P. L. Patterson, and H. Ramirez. 2003. *Statistical Strategy for Inventorying and Monitoring the Ecosystem Resources of the States of Jalisco and Colima at Multiple Scales and Resolution Levels*. RMRS-GTR-107, Rocky Mountain Research Station, USDA Forest Service, Fort Collins, Colorado, USA.

Stehman, S. V. 1992. Comparison of systematic and random sampling for estimating the accuracy of maps generated from remotely sensed data. *Photogrammetric Engineering and Remote Sensing* 58:1343–1350.

Stehman, S. V. 1996a. Estimating the Kappa coefficient and its variance under stratified random sampling, *Photogrammetric Engineering and Remote Sensing* 62:401–402.

Stehman, V. S. 1996b. Cost-effective, practical sampling strategies for accuracy assessment of large-area thematic maps. In *Spatial Accuracy Assessment in Natural Resources and Environmental Sciences: Second International Symposium*, May 21–23, Fort Collins, Colorado, pp. 485–492.

Stohlgren, T. J., G. W. Chong, M. A. Kalkhan, and L. D. Schell. 1997a. Rapid assessment of plant diversity patterns. A methodology for landscapes. *Environmental Monitoring and Assessment* 48:25–43.

Stohlgren, T. J., G. W. Chong, M. A. Kalkhan, and L. D. Schell. 1997b. Multi-scale sampling of plant diversity: Effects of minimum mapping unit size. *Ecological Application* 7:1064–1074.

Thompson, S. K. 1992. *Sampling*. New York: John Wiley & Sons.

Tryfos, P. 1996. *Sampling Methods for Applied Research: Text and Cases*. New York: John Wiley & Sons.

Walsh, T. A. and T. E. Burk. 1993. Calibration of satellite classifications of land area. *Remote Sens. Environ.* 46, pp. 281–290.

# 3

## Spatial Pattern and Correlation Statistics

### Scale

Scale represents the real world as translated onto a map; in other words, it is the relationship between distance on a map, image, air photo and the corresponding distance on Earth (Malczewski 1999). Scale is also the spatial or temporal measure of an object or a process (Turner and Gardner 1991), or level or degree of spatial resolution (Forman 1995). Components of scale include composition, structure, and function, which are all important ecological and environmental concepts.

The spatial or temporal dimension over which an object or process can be said to exist, as in, for example, a landscape, forest ecosystem or community. With geospatial information, a scale on a map, aerial photo, Landsat image can be represented by:

- Small scale—1: 100,000, 1:250,000, 1:500,000, 1:1,000,000 (state, country, world)
- Medium scale—1:10,000, 1:15,000, 1:24,000, 1:50,000 (stand, park, city, urban features)
- Large scale—1:100, 1:500, 1:1000, 1:5000 (trees, pond, streams, street, building, other)

In geostatistical application, we prefer to use *fine* and *coarse* scale to describe the variability within the landscape of forest, rangeland, vegetation, or the extent of the study area. We can define fine and coarse scale in these terms: Data can be used to better extrapolate the landscape from fine to coarse scales in terms of the area sizes. With respect the definition into spatial statistical, the term for fine scale is when developing a regression model through some form of Ordinary Least Square or stepwise regress analysis or logistic regression or generalized linear model (GLM), the outcome for describing the landscape is called fine scale variability and adding the residual values (different between the observed and estimated values) represent coarse scale, the outcome called for describing the full extent of the spatial variability is to combine fine and coarse scale. Both fine and coarse scales are used differently from the above definition in geographic description. We will be using both terms for the purpose of geostatistical spatial pattern analysis, data interpolation (kriging and cokriging, semivariogram), and regression model and map development.

Researchers and resource manager teams need to consider the fundamental issue of scale, which requires ensuring that the conclusion of the analysis does not depend on any arbitrary scale. Natural resource and landscape ecologists failed to do this for many years and for a long time characterized landscape elements with quantitative metrics, which depended on the scale at which they were measured.

## Spatial Sampling

(This section is cited and modified from http://www.wikipedia.org to reflect the objective of this book.) In geostatistical analysis often we use the term *spatial sampling* to relate geographical locations on landscape of forest and range lands, agricultural systems, water bodies, and so forth. This process involves determining a limited number of locations in a geospace (location based on X and Y coordinates) for faithfully measuring phenomena that are subject to dependency and heterogeneity. In addition, dependency suggests that since one location can predict the value of another location we do not need observations in both places. But heterogeneity suggests that this relation can change across space and therefore we cannot trust that an observed degree of dependency beyond a region may be small. Basic spatial sampling schemes include random, clustered, and systematic. These basic schemes can be applied at multiple levels in a designated spatial hierarchy (e.g., urban area, city, and neighborhood). It is also possible to exploit ancillary data, for example, using property values as a guide in a spatial sampling scheme to measure educational attainment and income. Spatial models such as autocorrelation statistics, regression, and interpolation (see later) can also dictate sampling.

### Errors in Spatial Analysis

The most common error problems in spatial analysis are including bias (error of measurement between actual values to their estimates), distortion, and outright errors in the conclusions reached by researchers or analysts using the data.

### Spatial Variability and Method of Prediction

Different things can be arranged on a map in different ways (landscape scale, plant distribution, forest structure, fuel loading, wetlands and depressional wetlands, roads, etc.). For example, the number, size, shape, and distance between patches of trees, can describe the pattern of forest patches. The spatial pattern exhibited by a map can also be described in terms of its overall texture, complexity, and other indicators.

---

## Spatial Pattern

Spatial pattern can be defined as how a certain population occupies a landscape (structure, placement, or arrangement of objects on Earth) and space between objects. In ecology it is often difficult to study the movement of individual animals directly, especially for small, numerous animals such as insects. Ecologists have therefore studied the spatial pattern of individuals of a particular species to infer the underlying behavioral rules that govern their movement (Greig-Smith 1952; Lloyd 1967; Kennedy 1972; Taylor 1986). The spatial heterogeneity that resulted from nonrandom interactions between individuals, from both inter- and intraspecific behavior, is intended to stabilize ecological systems (Hassell and May 1973). Spatiotemporal dynamic ecological models (Czaran and Bartha 1992), such as cellular automata (Hassell et al. 1991) and metapopulation models (Hanski and Gilpin 1991; Perry and Gonzalez-Andujar 1993; Perry 1994) increasingly use space explicitly to locate and move individuals within a two-dimensional coordinate system. Wiens (1989) stressed the importance of spatial scale and emphasized that an ecological process that operates in a certain way at one scale may not operate in the same way, or at all, at a different scale (Heads and Lawton 1983).

Some general geospatial considerations we need to take in terms of spatial patterns include:

Patterns may be recognized because of their arrangement, maybe in a line or by a clustering of points. In attempting to recognize and understand spatial patterns on maps or air photos, we need to ask the following questions:

- Is there an area that is denser with objects than others?
- Is there an area that has fewer or no objects than others?
- Are there clusters of objects?
- Is there a randomness or uniformity to the location of the objects?
- Does there seem to be a relationship between individual objects (is one object located where it is because of another)?

### Spatial Point Pattern

A spatial point pattern is a spatial pattern that is composed of closely arranged, somewhat organized, points. Examine the map of cities in the

United States in Figure 3.1 and try to pick out patterns. An example of spatial line patterns might be found on a map of roads or river networks. Look at the map of U.S. rivers in Figure 3.2 and note the patterns. In addition to the aforementioned questions, ask:

- Are the lines leading from one to another or leading from another feature?
- Are the lines connecting to each other in a meaningful manner?

USA Cities with Population larger than 15,000 People

**FIGURE 3.1**
Cities in the United States is a good example for spatial pattern, point data. (Adapted from the U.S. Census Web site.)

Spatial Line Pattern for Rivers in USA

**FIGURE 3.2**
Rivers are representative of spatial line patterns. (Adapted from the U.S. Census Web site.)

Last, we need to consider area patterns. A few things to look for in terms of interpretation of areal patterns are:

- Representative textures
- Colors and color transitions
- Areas that are adjacent or nearby one another

Look at the U.S. population density map in Figure 3.3 and note the patterns.
In general there are three forms of patterns: aggregate (cluster), regular, and random.

1. A *cluster pattern* is represented by a set of data and has one or more groups of points in clusters that cover large areas of landscape (Figure 3.4).
2. A *regular pattern* is regularly dispersed and it is implying the events are distributed more or less regularly over the landscape (Figure 3.4).
3. Random pattern implies the data set has no dominant trend toward clustering or dispersion (Figure 3.4).

Cressie (1993), Diggle (2003), Stoyan and Stoyan (1994), and Upton and Fingleton (1985) point out criteria used to determine if a data set is appropriate for any type of point pattern analysis:

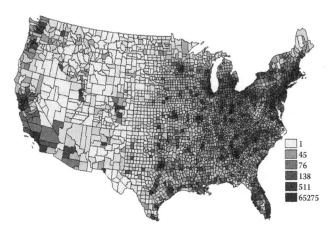

**USA Population Density by Counties (person per square mile):**
**Light color represents less population vs. Dark color more or dense**

**FIGURE 3.3**
U.S. counties for spatial areal pattern. (Adapted from the U.S. Census Web site.)

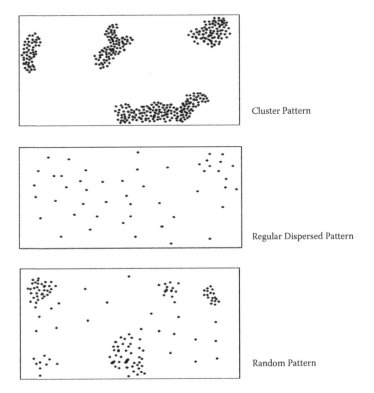

**FIGURE 3.4**
Representation of different spatial patterns (cluster, regular, and random).

- Spatial data must be mapped on a plane; both latitude and longitude coordinates are needed.
- The study area must be selected and determined prior to the analysis.
- Point data should not be a selected sample, but rather the entire set of data to be analyzed.
- There should be a one-to-one correspondence between objects in study area and events in pattern.
- Points must be true incidents with real spatial coordinates.

Cressie (1993) described several methods and algorithms that can be used to describe a pattern for a collection of points. These methods are cited as follows:

1. Quadrant count method
2. Kernel density estimation (K means)

3. Nearest neighbor distance
   a. G function
   b. F function
   c. K function

Although these techniques can be more descriptive and more accurate, it is also true that many of these other methods are more complicated and difficult to implement. Researchers in ecological studies found that quadrant count analysis is a relatively easy method to implement and it has provided several opportunities to apply basic mathematical and statistical concepts.

### Quadrant Count Method

The quadrant count method can be described as partitioning of the data set into $n$ equal-sized subregions; we will call these subregions quadrants (E. Tariq, 2004 Class Technical Report, School of Mines and Technology, South Dakota, USA). In each quadrant we will be counting the number of events that occur and it is the distribution of quadrant counts that will serve as our indicator of pattern. The choice of the quadrant size can greatly affect our analysis, where large quadrants produce a coarse description of the pattern. If the quadrant size is too small then many quadrants may contain only one event or they might not contain any events at all. We will use the rule of thumb that the area of a square is twice the expected frequency of points in a random distribution (i.e., $2(Area/n)$) where $n$ is the number of points in the sample size. After partitioning the data set into quadrants, the frequency distribution of the number of points per quadrant has been constructed. The mean and variance of the sample are then computed to calculate the variance to mean ratio (VMR). The following criteria based on VMR value can be used for interpretation as the VMR of a sample (or population):

If VMR > 1, the pattern is clustered; if VMR < 1, the pattern is regularly dispersed; or if VMR = 1, the pattern is random (see Figure 3.1).

## Linear Correlation Statistic

The *Pearson correlation coefficient* ($r$) indicates the strength and direction of a linear relationship between two random variables or measures the strength of the linear relationship between two variables. Its value is generally between −1 and 1, and is calculated as follows:

$$r = \frac{\sum XY - \frac{\sum X \sum Y}{N}}{\sqrt{\left(\sum Y^2 - \frac{\left(\sum Y\right)^2}{N}\right)\left(\sum X^2 - \frac{\left(\sum X\right)^2}{N}\right)}} \qquad (3.1)$$

The Pearson correlation coefficient value alone may not be sufficient to evaluate this relationship, especially in the case where the assumption of normality is incorrect.

### Case Study

The following is an example on a study to assess landscape scale under semi-arid environment within Grand Staircase-Escalante National Monument (GSENM), Utah. GSENM is a complex landscape of plant diversity and covers an area of about 2.1 million acres. Key biological parameters can be estimated using some form of multiscale sampling with multiphase design (i.e., double sampling) to provide unbiased estimates of plant species richness and diversity, invasive plants, rare habitat, and cryptobiotic crusts cover and their association with vegetation and soils characteristics. A total of 367 plots (0.1 ha) were established over the entire monument, and 19 vegetation cover types were found (Figure 3.5). The research was based on using field data, spatial information, and spatial statistics modeling and mapping.

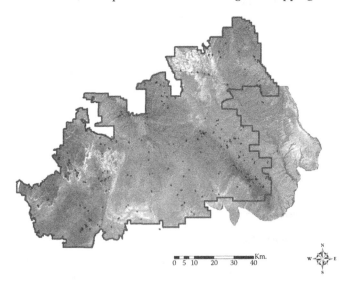

**FIGURE 3.5**
Overlay field plot locations with Landsat TM within GSENM area, Utah.

## Statistical Example

Following is an example using the Pearson correlation coefficient (r) to test for linear correlation between variables of interest. In S plus or R, use file name "gsenmplant" to run the test as:

```
> cor (gsenmplant [1:6] This will read from column 1 to 6:
```

Use these six variables: total plants (CtTot), percent covers of native total plants (CvNtv), total exotic (CtExo), cover total plants (CvTot), total native plants (CtNtv), and percent cover exotic (CvEXo).

```
> cor (gsenmplant [, 1:6])
```

|       | CtTot    | CvNtv     | CtExo    | CvTot   | CtNtv    | CvExo     |
|-------|----------|-----------|----------|---------|----------|-----------|
| CtTot | 1        | 0.29598   | 0.23019  | 0.26888 | 0.96190  | -0.04426  |
| CvNtv | 0.29598  | 1         | 0.06379  | 0.91639 | 0.290338 | -0.028131 |
| CtExo | 0.23019  | 0.06379   | 1        | 0.29347 | .00923   | 0.58149   |
| CvTot | 0.26888  | 0.91639   | 0.29347  | 1       | 0.20940  | 0.36999   |
| CtNtv | 0.96190  | 0.290338  | 0.00923  | 0.2094  | 1        | -0.17091  |
| CvExo | -0.04425 | -0.028131 | 0.581497 | 0.36999 | -0.17091 | 1         |

## Spatial Correlation Statistics

Spatial autocorrelation statistics measure the level of interdependence between the variables and the nature and strength of the interdependence (Fortin et al. 1989). Spatial autocorrelation may be classified as either positive or negative (zero means no spatial pattern). Positive spatial autocorrelation has all similar values appearing together, whereas negative spatial autocorrelation has dissimilar values appearing in close association. Furthermore, spatial autocorrelation is related to the scale of the data as a periodicity of elements is assessed. Negative spatial autocorrelation is more sensitive to changes in scale. In ecological geographic applications there is usually positive spatial autocorrelation.

The most commonly used measures for spatial autocorrelation in ecological, health, environmental, and geological studies are the Moran's *I* statistic (Moran 1948), Geary's *C* statistic (Geary 1954), and the spatial cross-correlation statistic (Fortin et al. 1989; Bonham et al. 1995; Reich et al. 1995; Kalkhan and Stohlgren 2000; Kalkhan et al. 2007). These are especially useful in studies of patchily distributed resources in time and space (e.g., water, nutrients; Bartell and Brenkert 1991) as determinants of vegetation patterns (Reich et al. 1995). Spatial autocorrelation and cross-correlation statistics can be used to evaluate spatially explicit information (e.g., information on the spatial pattern and scale between the variables of interest within landscape environments) on vegetation characteristics and structures (e.g., plant cover abundance, patterns of old-growth forest stands, and plant species richness patterns), soil parameters, and associated environmental characteristics (e.g., topographic, edaphic, and resource availability; Reich et al. 1995; Kalkhan and Stohlgren 2000; Kalkhan et al. 2007a, 2007b).

## Moran's *I* and Geary's *C*

Moran's *I* and Geary's *C* is global in the sense that they estimate the overall degree of spatial autocorrelation for a data set. The possibility of spatial heterogeneity suggests that the estimated degree of autocorrelation may vary significantly across geospace. *Local spatial autocorrelation statistics* provide estimates disaggregated to the level of the spatial analysis units, allowing assessment of the dependency relationships across space. *G statistics* compare neighborhoods to a global average and identify local regions of strong autocorrelation. A local version of the Moran's *I* and Geary's *C* statistics is also available.

*Moran's I* behaves like a Pearson correlation coefficient. Its value is generally between –1 and 1, but can sometimes exceed –1 or +1 (Legendre and Fortin 1989). Positive values indicate positive autocorrelation and vice versa. Moran's *I* is calculated as follows:

$$I(d) = \frac{\frac{1}{W} \sum_i \sum_{i \neq i} w_{ij}(z_i - \bar{z}) \cdot (z_j - \bar{z})}{\frac{1}{n} \sum_i (z_i - \bar{z})^2} \tag{3.2}$$

where $I(d)$ is the Moran coefficient for the distance class $d$, $z_i$'s are the values of the variable, and $i$ and $j$ vary from 1 to $n$. $w_{ij}$'s take the value 1 when the pair of location $(i,j)$ pertains to distance class $d$ and 0 otherwise. $W$ is the sum of the $w_{ij}$'s.

*Geary's C* is similar to a distance-type structure function. Values of $C$ smaller than 1 correspond to positive autocorrelation, and the greater the positive autocorrelation, the lower the value of $C$. Geary's $C$ is calculated as follows:

$$C(d) = \frac{(n-1) \sum_i \sum_{i \neq i} w_{ij}(z_i - z_j)^2}{2W \sum_i (z_i - \bar{z})^2} \tag{3.3}$$

where $C(d)$ is the Geary coefficient for the distance class $d$ and $z_i$'s and $w_{ij}$'s are as in Equation 3.2.

Both Moran's *I* and Geary's *C* can be tested for significance against their theoretical distribution (Sokal and Oden 1978; Cliff and Ord 1981; Legendre and Legendre 1984).

These tests indicate the degree of spatial association as reflected in the data set as a whole. They both necessitate the choice of a spatial weights matrix. Whereas Moran's *I* is based on cross-products to measure value association, Geary's *C* employs squared differences (Legendre and Fortin 1989). Moran's *I* behaves like a Pearson correlation coefficient. Its value is between –1 and +1 when scaled, otherwise it can exceed –1 or +1 (Legendre and Fortin 1989). Geary's *C* is similar to a distance-type structure function and is within the range of 0 to +2 (Sokal and Oden 1978). Values of *C* smaller than 1 correspond to a positive spatial autocorrelation, whereas values greater than 1 through 2 represent a negative spatial autocorrelation (Sokal and Oden 1978). The greater the positive autocorrelation, the lower is the value of *C*. Both Moran's *I* and Geary's *C* can be tested for significance against their theoretical distribution (Sokal and Oden 1978; Cliff and Ord 1981; Legendre and Fortin 1989).

### Cross-Correlation Statistic

The *cross-correlation statistic* ($I_{YZ}$; Czaplewski and Reich 1993) is used to test the null hypothesis of no spatial cross-correlation among all pairwise combinations of vegetation, forest, fire loading variables, topographic, and edaphic characteristics. Cliff and Ord (1981) showed that $I_{YZ}$ ranges from –1 to +1, although it can exceed these limits with certain types of spatial matrices. Czaplewski and Reich (1993) provided the following equation to calculate the cross-correlation statistic:

$$I_{YZ} = \left(\frac{1}{2W}\right) \frac{\sum_{i=1}^{n}\sum_{j=1}^{n} w_{ij}(y_i - \bar{y})(z_j - \bar{z})}{\sqrt{Var(y)Var(z)}} \tag{3.4}$$

where $w_{ij}$ is a scalar that quantifies the degree of spatial association or proximity between locations $i$ and $j$, or a 0–1 variable indicating that locations $i$ and $j$ are within some distance range of each other; $W$ is the sum of all $n^2$ values of $w_{ij}$; Var($y$) is the sample variance of $y_i$; and Var ($z$) is the sample variance of $z_i$.

### Inverse Distance Weighting (IDW)

Moran's *I* and Geary's *C* require measuring a spatial weights matrix that reflects the intensity of the geospatial relationship between observations in a neighborhood, for example, the distances between neighbors, the lengths of shared border, or whether they fall into a specified directional class such as west. These spatial autocorrelation statistics compare the spatial weights to the covariance relationship at pairs of locations. Spatial autocorrelation that is more positive than expected from random indicate the clustering of similar values across geospace, whereas significant negative spatial autocorrelation indicates that neighboring values are more dissimilar than expected by chance, suggesting a spatial pattern similar to a chess board.

In calculating the various test statistics, inverse distance weighting was used to describe the spatial proximity of the sample plots to one another. All tests were preformed at the 0.05 level of significance. Significant cross-correlation does not imply a true cross-correlation. The test statistic is testing for both auto- and cross-correlations simultaneously. If significant cross-correlations are observed additional testing may be required to ascertain the true nature of the relationship between variables. Much statistical software can be used for spatial autocorrelation and cross-correlations.

The *inverse distance weighting* (IDW) technique is one of the most basic procedures of spatial interpolation approaches and served as a basis of other spatial interpolation techniques. A focal point estimate at certain locations is mainly associated with its neighboring sample points. The spatial weights are first computed from the distances of sample points to an estimated point (i.e., an unsampled location) using a specific degree of polynomial (Isaaks and Srivastava 1989; Erxleben et al. 2002; Reich and Davis 2003a; Fortin and Dale 2005). The higher degrees of polynomial indicate that the closer neighboring points are more important and assigned more weighting values than the farther ones.

The estimated values at a given location $(\hat{Z}_0)$ can be described as:

$$\hat{Z}_0 = \frac{\sum_{i=1}^{n} \dfrac{Z_i}{d_i^p}}{\sum_{i=1}^{n} \dfrac{1}{d_i^p}}$$

(3.5)

where $Z_i$ is the observed value at a sample point $i$; $d_i$ is the distance between observation $i$ to a given location; $p$ is 0, 1, 2, 3, 4, ... (degree of the polynomial); and $n$ is the total number of neighbors used to estimate an unknown location.

To obtain a spatial weight for interpolate a given location $i$ (or an unsampled point $i$ of an object, maybe a Landsat pixel or GIS grid cell), a leave-one-out cross-validation procedure will be used to indicate a suitable number of nearest neighbors and degree of the polynomial that minimizes estimated variances.

Reich and Davis (2003a) described estimated IDW variance $(\hat{\sigma}_{i(IDW)}^{2*})$ associated with a predicted value at a given location as:

$$\hat{\sigma}_{i(IDW)}^{2*} = \left(\frac{n}{n-1}\right) \frac{\sum_{i=1}^{n}\left(\dfrac{Z^i}{Z_i^p} - \dfrac{\hat{Z}}{d_i^p}\right)^2}{\left(\sum_{i=1}^{n} \dfrac{1}{d_i^p}\right)^2}$$

(3.6)

where $Z_i$ is the observed value at a sample point $i$; $\hat{Z}_0$ is the estimated value at a given location; $d_i$ is the distance between observation $i$ to a given location; $P$ is 0, 1, 2, 3, ... (degree of the polynomial); and $n$ is the total number of neighbors estimating an unknown location.

### Statistical Example

Using the data "gsenmplant," we can develop weighted inverse distance, Moran's $I$, Gary's $C$, and cross-correlation statistics (bi-Moran $I$) to investigate the spatial pattern of plant diversity with GSENM landscape. Using R or S-plus, the following are steps for analyzing spatial patterns

### *1. Develop Inverse Distance Weighting*

```
gsenmplant.wt<-spwtdist (gsenmplant [, 26], gsenmplant [, 27],
a = 1, band = 0, binary = F, rescale = T)
[gsenmplant [,26, gsenmplant [,27]: reading x and y values from
your data;
a= 1 mean use inverse distance; band= 0: mean no restriction
on weight being set =0; binary = F, do not use binary weight;
and rescale = F= do not rescale]
Characteristics of distance matrix
Dimension: 359 (no. of observations)
Average distance between points: 52935.37
Distance range: 141171.27
Minimum distance between points: 1 (When rescale = T, min.
distance = 1)
Quartiles
First:    29696.01
Median:   50756.03
Third:    74741.25
Maximum distance between points: 141172.27
```

### *2. Develop Moran's I*

```
> morani(gsenmplant [,1],w=gsenmplant.wt)
UNDER NORMAL APPROXIMATION
Moran's I is = 0.175239
Mean of I is = -0.002793
St. Dev of I = 0.019705
Z-Value      = 9.034867
P-Value(2-side)    = 0
UNDER RANDOMIZATION ASSUMPTION
Moran's I is = 0.175239
Mean of I is = -0.002793
St. Dev of I = 0.019704
Z-Value      = 9.035289
```

```
P-Value(2-side)      = 0
> morani(gsenmplant[,2],w=gsenmplant.wt)
UNDER NORMAL APPROXIMATION
Moran's I is = 0.266679
Mean of I is = -0.002793
St. Dev of I = 0.019705
Z-Value      = 13.67533
P-Value(2-side)      = 0
UNDER RANDOMIZATION ASSUMPTION
Moran's I is = 0.266679
Mean of I is = -0.002793
St. Dev of I = 0.019687
Z-Value      = 13.687874
P-Value(2-side)      = 0
> morani(gsenmplant[,3],w=gsenmplant.wt)
UNDER NORMAL APPROXIMATION
Moran's I is = 0.185687
Mean of I is = -0.002793
St. Dev of I = 0.019705
Z-Value      = 9.565088
P-Value(2-side)      = 0
UNDER RANDOMIZATION ASSUMPTION
Moran's I is = 0.185687
Mean of I is = -0.002793
St. Dev of I = 0.019624
Z-Value      = 9.604603
P-Value(2-side)      = 0
> morani(gsenmplant[,4],w=gsenmplant.wt)
UNDER NORMAL APPROXIMATION
Moran's I is = 0.252206
Mean of I is = -0.002793
St. Dev of I = 0.019705
Z-Value      = 12.940844
P-Value(2-side)      = 0
UNDER RANDOMIZATION ASSUMPTION
Moran's I is = 0.252206
Mean of I is = -0.002793
St. Dev of I = 0.019668
Z-Value      = 12.96544
P-Value(2-side)      = 0
> morani(gsenmplant[,5],w=gsenmplant.wt)
UNDER NORMAL APPROXIMATION
Moran's I is = 0.198685
Mean of I is = -0.002793
St. Dev of I = 0.019705
Z-Value      = 10.224704
P-Value(2-side)      = 0
UNDER RANDOMIZATION ASSUMPTION
Moran's I is = 0.198685
Mean of I is = -0.002793
```

```
St. Dev of I  = 0.019712
Z-Value       = 10.220845
P-Value(2-side)   = 0
> morani(gsenmplant[,6],w=gsenmplant.wt)
UNDER NORMAL APPROXIMATION
Moran's I is  = 0.161463
Mean of I is  = -0.002793
St. Dev of I  = 0.019705
Z-Value       = 8.335773
P-Value(2-side)   = 0
UNDER RANDOMIZATION ASSUMPTION
Moran's I is  = 0.161463
Mean of I is  = -0.002793
St. Dev of I  = 0.019463
Z-Value       = 8.439426
P-Value(2-side)   = 0
```

### 3. Develop Geary's C

```
> gearyc(gsenmplant[,1],w=gsenmplant.wt)
UNDER NORMAL APPROXIMATION
Geary's C is  = 0.835525
Mean of C is  = 1
St. Dev of C  = 0.022053
Z-Value       = -7.45819
P-Value(2-side)   = 0
UNDER RANDOMIZATION ASSUMPTION
Geary's C is  = 0.835525
Mean of C is  = 1
St. Dev of C  = 0.02209
Z-Value       = -7.445673
P-Value(2-side)   = 0
> gearyc(gsenmplant[,2],w=gsenmplant.wt)
UNDER NORMAL APPROXIMATION
Geary's C is  = 0.756443
Mean of C is  = 1
St. Dev of C  = 0.022053
Z-Value       = -11.0442
P-Value(2-side)   = 0
UNDER RANDOMIZATION ASSUMPTION
Geary's C is  = 0.756443
Mean of C is  = 1
St. Dev of C  = 0.022769
Z-Value       = -10.696723
P-Value(2-side)   = 0
> gearyc(gsenmplant[,3],w=gsenmplant.wt)
UNDER NORMAL APPROXIMATION
Geary's C is  = 0.829481
Mean of C is  = 1
```

```
St. Dev of C = 0.022053
Z-Value       = -7.732258
P-Value(2-side)    = 0
UNDER RANDOMIZATION ASSUMPTION
Geary's C is = 0.829481
Mean of C is = 1
St. Dev of C = 0.025105
Z-Value       = -6.792322
P-Value(2-side)    = 0
> gearyc(gsenmplant[,4],w=gsenmplant.wt)
UNDER NORMAL APPROXIMATION
Geary's C is = 0.758464
Mean of C is = 1
St. Dev of C = 0.022053
Z-Value       = -10.952553
P-Value(2-side)    = 0
UNDER RANDOMIZATION ASSUMPTION
Geary's C is = 0.758464
Mean of C is = 1
St. Dev of C = 0.023511
Z-Value       = -10.273324
P-Value(2-side)    = 0
> gearyc(gsenmplant[,5],w=gsenmplant.wt)
UNDER NORMAL APPROXIMATION
Geary's C is = 0.810525
Mean of C is = 1
St. Dev of C = 0.022053
Z-Value       = -8.591822
P-Value(2-side)    = 0
UNDER RANDOMIZATION ASSUMPTION
Geary's C is = 0.810525
Mean of C is = 1
St. Dev of C = 0.021751
Z-Value       = -8.711202
P-Value(2-side)    = 0
> gearyc(gsenmplant[,6],w=gsenmplant.wt)
UNDER NORMAL APPROXIMATION
Geary's C is = 0.863476
Mean of C is = 1
St. Dev of C = 0.022053
Z-Value       = -6.190732
P-Value(2-side)    = 0
UNDER RANDOMIZATION ASSUMPTION
Geary's C is = 0.863476
Mean of C is = 1
St. Dev of C = 0.030236
Z-Value       = -4.515325
P-Value(2-side)     = 6e-006
```

### 4. Develop Bi-Moran's I

```
> bimorani(gsenmplant[,1], gsenmplant[,2],w=gsenmplant.wt)
UNDER RANDOMIZATION ASSUMPTION
Moran's I is = 0.144247
Mean of I is = -0.000827
St. Dev of I = 0.01458
Z-Value      = 9.949966
P-Value(2-side)     = 0
> bimorani(gsenmplant[,1], gsenmplant[,3],w=gsenmplant.wt)
UNDER RANDOMIZATION ASSUMPTION
Moran's I is = -0.029838
Mean of I is = -0.000643
St. Dev of I = 0.014337
Z-Value      = -2.036285
P-Value(2-side)     = 0.041722
> bimorani(gsenmplant[,1], gsenmplant[,4],w=gsenmplant.wt)
UNDER RANDOMIZATION ASSUMPTION
Moran's I is = 0.10979
Mean of I is = -0.000751
St. Dev of I = 0.014478
Z-Value      = 7.635247
P-Value(2-side)     = 0
> bimorani(gsenmplant[,1], gsenmplant[,5],w=gsenmplant.wt)
UNDER RANDOMIZATION ASSUMPTION
Moran's I is = 0.180523
Mean of I is = -0.002687
St. Dev of I = 0.019342
Z-Value      = 9.472035
P-Value(2-side)     = 0
> bimorani(gsenmplant[,1], gsenmplant[,6],w=gsenmplant.wt)
UNDER RANDOMIZATION ASSUMPTION
Moran's I is = -0.068093
Mean of I is = 0.000124
St. Dev of I = 0.014002
Z-Value      = -4.871905
P-Value(2-side)     = 1e-006
> bimorani(gsenmplant[,2], gsenmplant[,3],w=gsenmplant.wt)
UNDER RANDOMIZATION ASSUMPTION
Moran's I is = -0.022841
Mean of I is = -0.000178
St. Dev of I = 0.014009
Z-Value      = -1.617765
P-Value(2-side)     = 0.105713
> bimorani(gsenmplant[,2], gsenmplant[,4],w=gsenmplant.wt)
UNDER RANDOMIZATION ASSUMPTION
Moran's I is = 0.244221
Mean of I is = -0.00256
St. Dev of I = 0.01888
```

```
Z-Value        = 13.070851
P-Value(2-side)    = 0
> bimorani(gsenmplant[,2], gsenmplant[,5],w=gsenmplant.wt)
UNDER RANDOMIZATION ASSUMPTION
Moran's I is = 0.148349
Mean of I is = -0.000811
St. Dev of I = 0.014565
Z-Value        = 10.241005
P-Value(2-side)    = 0
> bimorani(gsenmplant[,2], gsenmplant[,6],w=gsenmplant.wt)
UNDER RANDOMIZATION ASSUMPTION
Moran's I is = -0.013761
Mean of I is = 7.9e-005
St. Dev of I = 0.013992
Z-Value        = -0.989127
P-Value(2-side)    = 0.322601
> bimorani(gsenmplant[,3], gsenmplant[,4],w=gsenmplant.wt)
UNDER RANDOMIZATION ASSUMPTION
Moran's I is = 0.040959
Mean of I is = -0.00082
St. Dev of I = 0.014555
Z-Value        = 2.870359
P-Value(2-side)    = 0.0041
> bimorani(gsenmplant[,3], gsenmplant[,5],w=gsenmplant.wt)
UNDER RANDOMIZATION ASSUMPTION
Moran's I is = -0.061583
Mean of I is = -2.6e-005
St. Dev of I = 0.01399
Z-Value        = -4.400275
P-Value(2-side)    = 1.1e-005
> bimorani(gsenmplant[,3], gsenmplant[,6],w=gsenmplant.wt)
UNDER RANDOMIZATION ASSUMPTION
Moran's I is = 0.134163
Mean of I is = -0.001624
St. Dev of I = 0.016121
Z-Value        = 8.422875
P-Value(2-side)    = 0
> bimorani(gsenmplant[,4], gsenmplant[,5],w=gsenmplant.wt)
UNDER RANDOMIZATION ASSUMPTION
Moran's I is = 0.106495
Mean of I is = -0.000585
St. Dev of I = 0.014297
Z-Value        = 7.489761
P-Value(2-side)    = 0
> bimorani(gsenmplant[,4], gsenmplant[,6],w=gsenmplant.wt)
UNDER RANDOMIZATION ASSUMPTION
Moran's I is = 0.051764
Mean of I is = -0.001034
St. Dev of I = 0.014884
Z-Value        = 3.547288
```

```
P-Value(2-side)     = 0.000389
> bimorani(gsenmplant[,5], gsenmplant[,6],w=gsenmplant.wt)
UNDER RANDOMIZATION ASSUMPTION
Moran's I is = -0.097047
Mean of I is = 0.000477
St. Dev of I = 0.014193
Z-Value      = -6.871503
P-Value(2-side)     = 0
```

# References

Abler, R., J. Adams, and P. Gould. 1971. *Spatial Organization—The Geographer's View of the World*. Englewood Cliffs, NJ: Prentice-Hall.

Anselin, L. 1995. Local indicators of spatial association—LISA. *Geographical Analysis* 27:93–115.

Bartell, S. M., and A. L. Brenkert. 1991. A spatial-temporal model of nitrogen dynamics in a deciduous forest watershed. In *Quantitative Methods in Landscape Ecology* (Ecological Studies Vol. 82), M. G. Turner and R. H. Gardner, eds., 379–398. New York: Springer-Verlag.

Benenson, I., and P. M. Torrens. 2004. *Geosimulation: Automata-Based Modeling of Urban Phenomena*. Chichester, UK: Wiley.

Bonham, C. D., and R. M. Reich. 1999. Influence of spatial autocorrelation on a fixed-effect model used to evaluate treatment of oil spills. *Applied Mathematics and Computation* 106:149–162.

Bonham, C. D., R. M. Reich, and K. K. Leader. 1995. Spatial cross-correlation of *Boutelua gracilis* with site factor. *Grassland Science* 41:196–201.

Cliff, A. D. and Ord, J. K., 1981. *Spatial Processes, Models and Applications*. Pion Ltd., London, England, pp. 21–45.

Cressie, N. A. 1993. *Statistics for Spatial Data* (rev. ed.). New York: John Wiley & Sons.

Czárán, T., and S. Bartha. 1992. Spatiotemporal dynamicmodels of plant populations and communities. *Trends in Ecology and Evolution* 7:38–42.

Czaplewski, R. L., and R. M. Reich. 1993. Expected Value and Variance of Moran's Bivariate Spatial Autocorrelation Statistic under Permutation. USDA Forest Service Research Paper RM-309. Fort Collins, Colorado, pp. 1–13.

Dale, M. R. T. 1999. *Spatial Pattern Analysis in Plant Ecology*. Cambridge: Cambridge University Press.

Davis, J. 1986. *Statistics and Data Analysis in Geology*. Toronto: John Wiley & Sons.

Diggle, P. J. 2003. *Statistical Analysis of Spatial Point Patterns*, 2nd ed. London: Arnold Publishers.

Erxleben J., K. Elder, and R. Davis. 2002. Comparison of spatial interpolation methods for estimating snow distribution in the Colorado Rocky Mountains. *Hydrological Process* 16:3627–3649.

Forman, R. T. T. 1995. *Land Mosaics: The Ecology of Landscapes and Regions*. Cambridge: Cambridge University Press.

Fortin, M. J., Drapeau, P., Legendre, P. 1989. Spatial auto-correlation and sampling design in plant ecology. *Vegetatio* 83, 209–222.

Fortin, M. J., and M. Dale. 2005. *Spatial Analysis: A Guide for Ecologists*. Cambridge: Cambridge University Press.

Fotheringham, A. S., C. Brunsdon, and M. Charlton. 2000. *Quantitative Geography: Perspectives on Spatial Data Analysis*. London: Sage.

Fotheringham, A. S., and M. E. O'Kelly. 1989. *Spatial Interaction Models: Formulations and Applications*. Dordrecht: Kluwer Academic.

Fotheringham, A. S., and P. A. Rogerson. 1993. GIS and spatial analytical problems. *International Journal of Geographical Information Systems* 7:3–19.

Geary, R. C. 1954. The contiguity ratio and statistical mapping. *The Incorporated Statistician* 5:115–145.

Gergel, S. E., and M. G. Turner (eds.). 2002. *Learning Landscape Ecology: A Practical Guide to Concepts and Techniques*. New York: Springer-Verlag.

Goodchild, M. F. 1987. A spatial analytical perspective on geographical information systems. *International Journal of Geographical Information Systems* 1:327–344.

Greig-Smith, P. 1952. The use of random and contiguous quadrats in the study of the structure of plant communities. *Annals of Botany* 16:293–316.

Grunwald, S. 2006. *Environmental Soil Landscape Modeling*. New York: CRC Press.

Haining, R. 1990. *Spatial Data Analysis in the Social and Environmental Sciences*. Cambridge: Cambridge University Press.

Hanski, I., and M. Gilpin. 1991. Metapopulation dynamics: brief history and conceptual domain. *Biological Journal of the Linnean Society* 42:3–16.

Hassell, M. P., and R. M. May. 1973. Stability in insect hostparasite models. Journal of Animal Ecology 42:693–726.

Isaaks, E. H., and R. M. Srivastava. 1989. *An Introduction to Applied Geostatistics*. New York: Oxford University Press.

Kalkhan, M. A., R. M. Reich, and R. L. Sanford, Jr. 1993. Spatial analysis of canopy openings in a primary neotropical lowland forest. In *ACSM/ASPRS Annual Convention and Exposition*, ASPRS Technical Papers, February 16–19, New Orleans, pp. 175–182.

Kalkhan, M. A., E. J. Stafford, and T. J. Stohlgren. 2007a. Rapid plant diversity assessment using a pixel nested plot design: A case study in Beaver Meadows, Rocky Mountain national Park, Colorado, USA. *Diversity and Distribution* 13:379–388.

Kalkhan, M. A., E. J. Stafford, P. J. Woodly, and T. J. Stohlgren. 2007b. Exotic plant species invasion and associated abiotic variables in Rocky Mountain National Park, Colorado, USA. *Journal of Applied Soil Ecology* 37:622–634.

Kalkhan, M. A., and T. J. Stohlgren. 2000. Using multi-scale sampling and spatial cross-correlation to investigate patterns of plant species richness. *Environmental Monitoring and Assessment* 64:591–605.

Kanevski, M., and M. Maignom. 2004. Analysis and Modeling of Spatial Environmental Data. New York: Marcel and Dekker.

Kennedy, J. S. 1972. The emergence of behaviour. *Journal of the Australian Entomological Society* 11:168–176.

Legendre, P. and Fortin, M. J., 1989. Spatial analysis and ecological modelling. *Vegetatio* 80, 107–138.

Legendre, P. and V. Legendre. 1984. Postglacial dispersal of freshwater fishes in the Québec peninsula. Can J. Fish. Aquat. Sci. 41: 1781–1802

Lloyd, M. 1967. 'Mean crowding.' *Journal of Animal Ecology* 36:1–30.

MacEachren, A. M., and D. R. F. Taylor (eds.). 1994. *Visualization in Modern Cartography.* Oxford: Pergamon Press.

Malczewski, J. 1999. *GIS and Multicriteria Decision Analysis.* New York: John Wiley & Sons.

Miller, H. J. 2004. Tobler's first law and spatial analysis. *Annals of the Association of American Geographers* 94, 284–289.

Miller, H. J., and J. Han (eds.). 2001. *Geographic Data Mining and Knowledge Discovery.* Boca Raton, FL: Taylor & Francis.

Moran, P. A. P. 1948. The interpretation of statistical maps. *Royal Statistical Society, Serial B* 10:243–351.

O'Sullivan, D., and D. Unwin. 2002. *Geographic Information Analysis.* Hoboken, NJ: Wiley.

Parker, D. C., S. M. Manson, M. A. Janssen, M. J. Hoffmann, and P. Deadman. 2003. Multi-agent systems for the simulation of land-use and land-cover change: A review. *Annals of the Association of American Geographers* 93:314–337.

Perry, J. N. 1994. Chaotic dynamics can generate Taylor's power law. *Proceedings of the Royal Society of London series B* 257:221–226.

Perry, J. N., and J. L. Gonza'lez-Andujar. 1993. A metapopulation neighbourhood model of an annual plant with a seedbank. *Journal of Ecology* 81:453–463.

Reich, R. M., R. L. Czaplewski, and W. A. Bechtold. 1995. Spatial cross-correlation of undisturbed natural shortleaf pine stands in northern Georgia. *Environmental and Ecological Statistics* 1: 201–217.

Reich, R. M., and R. A. Davis. 1998. On-line spatial library for the S-Plus statistical software package. Colorado State University, Fort Collins, CO. (Available at http://www.warnercnr.colostate.edu/~robin/)

Reich, R. M., and R. A. Davis. 2003a. *Quantitative Spatial Analysis: Course Note for NR/ST523.* Fort Collins, CO: Colorado State University.

Reich, R. M., and R. A. Davis. 2003b. *Quantitative Spatial Analysis: Course Note for NR512.* Fort Collins, CO: Colorado State University.

Ripley, B. D. 1981. *Spatial Statistic.* New York: John Wiley & Sons.

Sokal, R. R., and N. L. Oden. 1978. Spatial autocorrelation in biology. 1. Methodology. *Biological Journal of the Linnean Society* 10:199–228.

Sokal, R. R., and J. D. Thomson. 1987. Applications of spatial autocorrelation in ecology. In *Developments in Numerical Ecology*, NATO ASI Series, Vol. G14, P. Legendre and L. Legendre (eds.). Berlin: Springer-Verlag.

Stoyan, D., and H. Stoyan. 1994. *Fractals, Random Shapes and Point Fields* (Methods of Geometrical Statistics). New York: John Wiley & Sons.

Taylor, L. R. 1986. Synoptic dynamics, migration and the Rothamsted Insect Survey. *Journal of Animal Ecology* 55: 1–38.

Turner, M. G., and R. H. Gardner (eds.). 1991. *Quantitative Methods in Landscape Ecology.* Chichester, UK: Springer-Verlag.

Upton, G. J. G., and B. Fingleton. 1985. *Spatial Data Analysis by Example* (Point Pattern and Quantitative Data), Vol. 1. Chichester, UK: John Wiley & Sons.

Wang, F. 2006. *Quantitative Methods and Applications in GIS.* New York: Taylor & Francis.

White, R., and G. Engelen 1997. Cellular automata as the basis of integrated dynamic regional modeling. *Environment and Planning B: Planning and Design* 24, 235–246.

Wiens, J. A. 1989. Spatial scaling in ecology. *Functional Ecology* 3:385–397.

# 4

## Geospatial Analysis and Modeling–Mapping

Ecological variables (e.g., vegetation, soil, and hydrologic conditions) and environmental characteristics measured in the field are important elements for modeling the fine-scale variability within landscape characteristics. These variables also help to produce reliable full-coverage thematic maps of the landscape. Gown et al. (1994) stated that many spatial data sets (i.e., remotely sensed data) provide reliable information for macroscale ecological monitoring, but they fall short in providing the precision required by more refined ecosystem resource models. Spatial statistics provide a means to develop spatial models that can be used to correlate coarse scale geographical data (i.e., remotely sensed imagery, topographic variables) with field measurements of biotic variables, such as those measured in a depressional wetland. If a satellite image is geographically referenced to a base map, one can overlay the location of field plots on the image to obtain reflectance values associated with each of the field plots. Then, if the field data are spatially correlated with reflectance from the remotely sensed image, it is possible to develop a model describing this spatial continuity (Cliff and Ord 1981).

## Stepwise Regression

When dealing with multiple variables (more than two independent variables) we may not able to evaluate all possible regression models for variables of interest in order to select the best fitting model during the process of developing a multiple regression model. There are numerous methods for selecting a subset of predictor variables in regression (see Myers 1986; Miller 1990). The following describes the ones most often used:

1. Forward selection begins with no centers in the network (an empty subset). At each step the center is added that most decreases the objective function; in other words, the independent variable that gives the largest reduction of the residual sum of squares.
2. Backward elimination begins with all candidate centers in the network (complete set). At each step the independent variable that gives the smallest increase in the residual sum of squares is dropped.

3. Efroymson's stepwise method is like forward selection, except that when each new variable is added to the subset, partial correlations are considered to see if any of the variables in the subset should now be dropped.

4. Leaps and bounds (Furnival and Wilson 1974) is an algorithm for determining the subset of centers that minimizes the objective function. This optimal subset can be found without examining all possible subsets, but the algorithm is practical only up to 30 to 50 candidate centers.

### Statistical Example

Using a stepwise procedure to identify which variable to be included in the regression model, we have data to predict total plants (360 plots) from Grand Staircase Escalante National Monument (GSENM), Utah. The independent variables are band (band1, band2, band3, band4, band5, band6l, band6h, band7, band8); tassel cap (tasl1, tasl2, tasl3, tasl4, tasl5, tasl6); and the topographic parameters are elevation (elv), slope (slp), aspect (absasp), landform. We use stepwise as:

```
gsenmplant[1,]  (reading row 1)

CtTot band1 band2 band3 band4 band5 band6l band6h band7 band8
25     101   96    113   71    157   163    208    121   71

tasl1    tasl2    tasl3    tasl4    tasl5    tasl6    elv
248      -68      -57      20       -37      -20      1599.289
slp      absasp   landform
29.9178  178.8621 -0.3047608

stepwise(gsenmplant[,c(7:19)],gsenmplant[,1])

$rss: Residual sum square for the significant variable (T)
[1] 27775.60 27583.78 27256.71

$size:
[1] 1 2 3

$which:
        band1 band2 band3 band4 band5 band6l band6h band7 band8 tasl1 tasl2
1(+ 7) F    F     F     F     F     F      T      F     F     F     F
2(+13) F    F     F     F     F     F      T      F     F     F     F
3(+12) F    F     F     F     F     F      T      F     F     F     F

        tasl3    tasl4

1(+ 7)    F       F
2(+13)    F       T
3(+12)    T       T
```

```
$f.stat: F- Statistics

[1] 37.040572 2.475777 4.259813

$method:

[1] «efroymson»
```

## Ordinary Least Squares (OLS)

After we used the stepwise procedure, one of the options we can implement is to use ordinary least squares (OLS). OLS is a mathematical optimization technique, which when given a series of measured data, attempts to find a function that closely approximates the data (a best fit). It attempts to minimize the sum of the squares of the ordinate differences (called residuals) between points generated by the function and corresponding points in the data. Specifically, it is called least mean squares (LMS) when the number of measured data is one and the gradient descent method is used to minimize the squared residual. LMS is known to minimize the expectation of the squared residual, with the smallest operations (per iteration). But it requires a large number of iterations to converge.

Kalkhan et al. (2000, 2001, 2004) developed geospatial–statistical models and maps for describing fine- and coarse-scale spatial variability to forecast plant diversity, vegetation–soil characteristics, cryptobiotic crust cover, invasive species, and fire fuel loading parameters within Rocky Mountain regions. All models were developed and modified based on Reich et al. (1999), in which they described a model based on the process using stepwise regression, trend surface analysis of geographical variables (e.g., elevation, slope, aspect, and landform), and measures of local taxa to evaluate coarse-scale spatial variability. The spatial statistical analyses in the proposed research are similar and will be accomplished using S-Plus (MathSoft 2000), as defined here:

$$\Phi_0 = \sum_{i+\ j \leq p}^{p}\sum^{p} \beta_{ij} \ x_{10}^i \ x_{20}^j + \sum_{K=1}^{q} \gamma_k \ y_{k0} + \eta_0 \tag{4.1}$$

where $\beta_{ij}$ is the regression coefficients associated with the trend surface component of the model, $y_{k0}$ is the regression coefficients associated with the $q$ auxiliary variables, $y_{k0}$, is available as a coverage in the geographic information system (GIS) database, and $\eta_0$ is the error term, which may or may not be spatially correlated with its neighbors (Kallas 1997; Metzger

1997). The least squares method fits a continuous, univariate response as a linear function of the predicted variable. This trend surface model represents continuous first-order spatial variation. Akaike's information criteria (AIC; Brockwell and Davis 1991; Akaike 1997) will be used as a guide in selecting the number of model parameters to include in the regression model where

$$\text{AIC} = -2 \text{ (max log likelihood)} + 2 \text{ (number of parameters)} \qquad (4.2)$$

When using maximum likelihood as a criterion for selecting between models of different orders, there is the possibility of finding another model with equal or greater likelihood by increasing the number of parameters (Metzger 1997). Therefore, the AIC allows for a penalty for each increase in the number of parameters. Using this criterion, a model with a smaller AIC will be considered to have a better fit. Although the model was kept as simplistic as possible, a more complex model could be used if the situation warrants it. The spatial statistical analysis, including spatial autocorrelation and cross-correlation, is then conducted ($\alpha = 0.05$).

The result from stepwise procedure identified band6h, tasl3, and tasl4 to be included in the model. Now, we can use the OLS model to develop trend surface (coarse-scale variability) for prediction total plant within GSENM, Utah. The fitted model is described as:

```
gsenmplant.ols<-ols(gsenmplant[,1],gsenmplant[,c(13,18,19)],w=
gsenmplant.wt)
```

```
Residual Standard Error = 8.7624, Multiple R-Square = 0.1109
N = 359, F-statistic = 14.7642 on 3 and 355 df, p-value = 0
```

|           | Coeffiicent | Std Errors | T-stat   | P-value |
|-----------|-------------|------------|----------|---------|
| Intercept | 71.8834     | 6.9618     | 10.3253  | 0.0000  |
| band6h    | -0.1670     | 0.0314     | -5.3177  | 0.0000  |
| tasl3     | 0.0674      | 0.0326     | 2.0639   | 0.0398  |
| tasl4     | -0.1417     | 0.0583     | -2.4323  | 0.0155  |

```
Log(like)          = -1286.5859
AIC                = 2581.1719
AICC               = 2581.2849
Schwartz           = 2596.7052
Moran's I (res)    = 0.1183
Mean of I          = -0.1968
Std Dev of I       = 0.1906
Z-value of I       = 1.6535
P-value(2-side)    = 0.0982
Lagrange Mult      = 34.6326
P-value(2-side)    = 0
```

Our conclusion based on the aforementioned OLS result that band6h, tasel3, and tasl4 which explained by 1% of the variability in predicting total plant. All variables are significant at the α level of 0.05 of significance. The AIC and AICc are lower for this model than the model use only band6h. The residual plot (not shown here) still suggests the presence of spatially correlated residuals, since Moran's $I$ test for the residual is not significant. However, examining the value from the Lagrange multiplier test, which is similar to chi-square statistics and used for testing the spatial autocorrelation of the residual), thus, suggest the residuals are spatially correlated. One may ask how this can happen; the answer is the two tests disagree with one another.

Once a spatial or temporal dependency is established for a given variable, this information can be used to interpolate values for points not measured (Robertson 1987). If these variables are spatially correlated with the variable of interest, this information can be used to improve estimates (Isaaks and Srivastava 1989). The use of auxiliary information in spatial prediction is referred to as cokriging. The usefulness of auxiliary information is enhanced by the fact that the variable of interest is generally undersampled (Isaaks and Srivastava 1989). Describing the spatial continuity is a concern in geostatistics modeling and mapping, especially dealing with research in landscape and ecological modeling, plant diversity, forest fire behavior, fuel variability, wildlife habitat, and urban and population growth. The most common statistical tool is the variogram. The variogram is the key function in geostatistical modeling and mapping applications and it can be used to fit a model of the spatial/temporal correlation of the observed phenomenon (i.e., distribution of tress, wildlife, presence and absence of plant, invasive community, disease and insect's intensity, forest and vegetation mortality, and landuse/landcover change).

## Variogram and Kriging

The variogram model is used to define the weights of the kriging function. Matheron (1962), Cliff and Ord (1980), and Cressie (1993) noticed that the experimental variogram is an empirical estimate of the covariance of a Gaussian process. As such, it may not be positive definite and hence not directly usable in kriging without constraints or further processing. This explains why only a limited number of variogram models are used like, such as the linear, the spherical, the Gaussian and the exponential.

When a variogram is used to describe the correlation of different variables it is called cross-variogram. Cross-variograms are used in cokriging.

Should the variable be binary or represent classes of values, one is then talking about indicator variograms. Indicator variograms are used in indicator kriging.

The semivariogram (usually termed variogram) is also a distance-type structure function (Matheron 1962). The lower the value of the variogram, the higher is the positive autocovariation. The variogram is analogous to a measure of variance and contrary to Geary's $C$; the value of the variogram depends on the units of the variable $z$.

The variogram is calculated as follows:

$$\gamma(d) = \frac{\displaystyle\sum_{i}\sum_{i \neq i} w_{ij}(z_i - z_j)^2}{2W} \tag{4.3}$$

where $\gamma(d)$ is the value of the variogram for distance class $d$, and $z_i$ and $w_{ij}$ are calculated as for Equation 3.4.

It is a univariate method limited to quantitative variables, allowing analyzing phenomena that occur in one, two, or three physical dimensions. Although, the data does not necessarily need to be stationary (absence of a spatial trend), the variogram computation requires their pseudostationarity, that is, that the dominant autocovariation is the same over the area of study. This is known as the intrinsic hypothesis.

The variogram, when calculated as in Equation 4.3, is called the *experimental variogram* since it is computed from the observed data. This variogram can be used for the description of spatial structures. A major feature of the variogram function is that it can be used as an input for the kriging method of interpolation (Matheron 1962; David 1977). In the kriging estimation procedure, a second type of variogram is required: the *theoretical variogram*. This variogram is based on a mathematical model, fitted to the experimental variogram. There are several models for theoretical variograms, and the most common are linear, spherical, exponential, and Gaussian. These are based on three parameters: the nugget, the sill, and the range. The nugget is the value of the variogram for distance 0 and is often assimilated to the sampling variance. The range is the distance from which the variogram ceases to increase; it defines the spatial domain of influence of the data. The sill is the value of the variogram when the range is reached. It is the maximum variability when there is no spatial autocorrelation left.

The linear model assumes a constant increase of the variance with the distance and hence there is neither range nor sill; it is the case of the spatial gradient. The three other models assume an increase of variability with the distance for small spatial scales and a constant or nearly constant variability for higher spatial scales. These three models differ in their shape, mostly near the origin.

The spherical variogram is given by the following equation:

$$\begin{cases} 0 < d < a \Rightarrow \gamma^* = c\left(1.5\frac{d}{a} - 0.5\frac{d^3}{a^3}\right) + c_0 \\ d \geq a \Rightarrow \gamma^* = c + c_0 \\ d = 0 \Rightarrow \gamma^* = 0 \end{cases} \tag{4.4}$$

where $\gamma^*$ is the value of the theoretical variogram, $d$ the distance between observations, $c_0$ the nugget, $c+c_0$ the sill, and $a$ the range.

Identification of the parameters $a$, $c$, and $c_0$ is done by nonlinear fitting of the theoretical variogram to the experimental one.

After selecting the best variogram model, we can explore kriging for modeling fine-scale variability of a landscape scale. Kriging is a method of interpolation named after a South African mining engineer, D. G. Krige, who developed the technique in an attempt to more accurately predict ore reserves. Over the past several decades kriging has become a fundamental tool in the field of geostatistics. Kriging is based on the assumption that the parameter being interpolated can be treated as a regionalized variable. A regionalized variable is intermediate between a truly random variable and a completely deterministic variable in that it varies in a continuous manner from one location to the next and therefore points that are near each other have a certain degree of spatial correlation, but points that are widely separated are statistically independent (Davis 1986). Kriging is a set of linear regression routines that minimizes estimation variance from a predefined covariance model. There are three techniques to perform kriging:

1. Ordinary kriging
2. Simple kriging
3. Universal kriging

### Ordinary Kriging

Ordinary kriging (OK) is a geostatistical approach to modeling. Instead of weighting nearby data points by some power of their inverted distance, ordinary kriging relies on the spatial correlation structure of the data to determine the weighting values. This is a more rigorous approach to modeling, as correlation between data points determines the estimated value at an unsampled point. The concept of spatial correlation and how to measure and develop a model is explained previously in this chapter. Furthermore, ordinary kriging makes the assumption of normality among the data points. When performing ordinary kriging or indicator kriging, semivariogram values for any distance may be required. Therefore, a model must be fit to the semivariogram values to provide a semivariogram value for any distance. To develop ordinary kriging construct a variogram from the scatter point set to be interpolated. A variogram consists of two parts: an experimental variogram and a model variogram (see Figure 4.1). Suppose that the value to be interpolated is referred to as *f*. The experimental variogram is found by calculating the variance (*g*) of each point in the set with respect to each of the other points and plotting the variances versus distance (*h*) between the points. Several formulae can be used to compute the variance, but it is typically computed as one-half the differences in f squared.

$$F(x,y) = \sum_{i=1}^{n} w_i f_i \text{ (Cited from Shepard 1968)}$$

where *n* is the number of scatter points in the set, $f_i$ is the values of the scatter points, and $w_i$ is the weight assigned to each scatter point. This equation is essentially the same as the equation used for inverse distance weighted interpolation (Equation 3.6) except that rather than using weights based on an arbitrary function of distance, the weights used in kriging are based on the model variogram. An important feature of kriging is that the variogram can be used to calculate the expected error of estimation at each interpolation point since the estimation error is a function of the distance to surrounding scatter points.

### Simple Kriging

Deutsch and Journel (1992) stated that simple kriging is similar to ordinary kriging and the weights do not sum to unity. Simple kriging uses the average of the entire data set, whereas ordinary kriging uses a local average (the average of the scatter points in the kriging subset for a particular interpolation

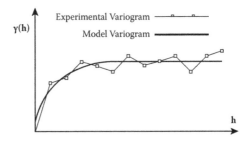

**FIGURE 4.1**
Experimental and model variograms used to construct ordinary kriging.

point). As a result, simple kriging can be less accurate than ordinary kriging, but it generally produces a result that is smoother and more aesthetically pleasing (Rich and Davis 2003).

## Universal Kriging

One of the assumptions made in kriging is that the data being estimated are stationary (Deutsch and Journel 1992). That is, as you move from one region to the next in the scatter point set, the average value of the scatter points is relatively constant. Whenever there is a significant spatial trend in the data values, such as a sloping surface or a localized flat region, this assumption is violated. In such cases, the stationary condition can be temporarily imposed on the data by use of a drift term. The drift is a simple polynomial function that models the average value of the scatter points. The residual is the difference between the drift and the actual values of the scatter points. Since the residuals should be stationary, kriging is performed on the residuals and the interpolated residuals are added to the drift to compute the estimated values. Using a drift in this fashion is often called universal kriging.

## Developing Variogram Model and Kriging to Predict Plant Diversity at GSENM, Utah

Using S-plus or R, we can develop variogram model as:

```
> gsenmplant.lst<-list(x=gsenmplant[,26],y=gsenmplant[,27],z=
gsenmplant.ols$resid, xl=369931,xu=523841,yl=4095242,yu=4207462)

> gsenmplant.lst
$x:
1        2        3        4        5        6       7
472457 404961   453725   454697   454661   406375  406251
8        9        10       11       12
402447 463428   435709   438236   421738

343      344      345      346      347      348      349      350      351
413559 453329 465161 417333 415099 414582 417221 414796 416468
352      353      354      355      356      357      358      359
413350 412098 420631 416828 416613 414686 412852 412519

$y:
1              2              3              4              5
4132051        4114792        4115716        4116885        4117024
6              7              8              9              10
4113027        4113094        4115913        4119874        4160172
```

| 351 | 352 | 353 | 354 | 355 |
|-----|-----|-----|-----|-----|
| 4129320 | 4141182 | 4148707 | 4146673 | 4123708 |
| 356 | 357 | 358 | 359 | |
| 4143738 | 4145543 | 4145769 | 4148976 | |

$z:

| 1 | 2 | 3 | 4 |
|---|---|---|---|
| -5.474044 | 8.172175 | 0.8237526 | 1.749206 |
| 5 | 6 | 7 | 8 |
| -0.1654438 | -5.58482 | -4.211589 | 10.19842 |

| 351 | 352 | 353 | 354 | 355 |
|-----|-----|-----|-----|-----|
| 15.22214 | -5.379745 | -4.816516 | -14.50109 | 3.726672 |
| 356 | 357 | 358 | 359 | |
| -10.90257 | -12.98074 | -14.41042 | -5.105246 | |

$xl:
[1] 369931
$xu:
[1] 523841
$yl:
[1] 4095242
$yu:
[1] 4207462

```
> summary(gsenmplant.ols$resid)
 Min. 1st Qu. Median Mean 3rd Qu. Max.
 -19.37 -6.116 0.08278 -1.732e-016 5.117 26.79
> par(mfrow=c(2,1))
> hist(gsenmplant.ols$resid)
> title(«Residuals»)
> plot(gsenmplant.ols$resid)
> title(«Residuals»)
> gsenmplant.var<-variogrm(gsenmplant[,26],gsenmplant[,27],gsenm
plant.ols$resid,nint=15,iso = T, theta = 0,dtheta =0,  dmax = 0)
```

In reading gsenmplnat: x [,26 and y[,27] from the file, nint = 15 is the number of bins to use in computing the sample variogram; iso = T is used to construct an all-directional variogram(default). The parameter theta and dtheta are used in computing directional variogram (iso = F); theta is the angle and dtheta is the tolerance. For instance, if theta = 20 and dtheta = 6 sample points separated by an angle of 20 degree plus or minus 6 degrees will be considered nearest neighbor when computing the sample variogram. The parameter dmax is the upper limit on the distance used in computing the directional variogram. If damx = 0, the sample variogram will be computed over the whole range of possible distance (see Reich and Davis 1998).

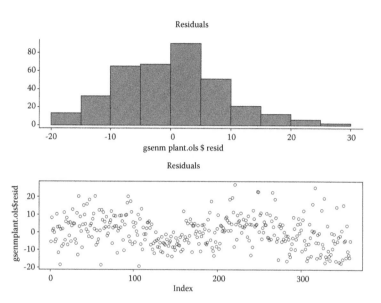

**FIGURE 4.2**
Histogram and distribution of residuals observed for plant species richness within GSENM, Utah.

```
> gsenmplant.lin<-fitvar(gsenmplant.var,0,80,65000,model="lin"
,wt=T)
Least Squares Estimate
Nugget        = 76.8145
Slope         = -2.3e-005
Log(like)     = -94.5822
AIC           = 193.1645
AICC          = 194.4978
Schwartz      = 194.1343
> gsenmplant.gau<-fitvar(gsenmplant.var,0,80,65000,model="gau"
,wt=T)
Least Squares Estimate
Nugget        = 56.817
Sill          = 77.2789
Range         = 2298.552826
alpha         = 0.735221
s.e.          = 8.790842
Log(like)     = -100.0144
AIC           = 206.0287
AICC          = 209.0287
Schwartz      = 207.4835
> gsenmplant.exp<-fitvar(gsenmplant.var,0,80,65000,model="exp"
,wt=T)
```

```
Least Squares Estimate
Nugget       = 56.817
Sill         = 77.278903
Range        = 572.945083
alpha        = 0.735221
s.e.         = 8.790842
Log(like)    = -100.0144
AIC          = 206.0287
AICC         = 209.0287
Schwartz     = 207.4835
> gsenmplant.sph<-fitvar(gsenmplant.var,0,80,65000,model="exp"
,wt=T)
Least Squares Estimate
Nugget       = 56.817
Sill         = 77.278903
Range        = 572.945083
alpha        = 0.735221
s.e.         = 8.790842
Log(like)    = -100.0144
AIC          = 206.0287
AICC         = 209.0287
Schwartz     = 207.4835
> x<-seq(0,120000, 100)
> lines(gauvar(x,gsenmplant.gau),lty=1,lwd=3)
> lines(expvar(x,gsenmplant.exp),lty=3,lwd=3)
> lines(sphervar(x,gsenmplant.sph),lty=4,lwd=3)
> legend(.3,.2,legend=c("Gaussian","Exponential","Spherical",l
ty=c(1,3,4)))
> title("Variogram Models, weighted")
```

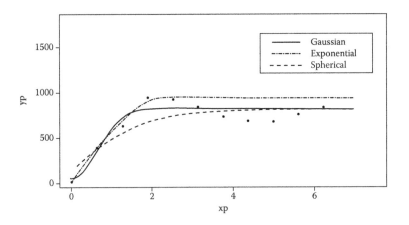

**FIGURE 4.3**
Example represents weighted variogram models (Gaussian, exponential, and spherical).

### Model Cross-Validation

```
> gsenmplant.val<-crossval(gsenmplant.lst,gsenmplant.gau,15)
> var(gsenmplant.val$resid)^.5
          [,1]
[1,] 8.000361
> 1-var(gsenmplant.val$resid)/var(gsenmplant[,1])
          [,1]
[1,] 0.2525778
> (523841 - 369939) / 1000 (calculate map resolution)
[1] 153.902
>gsenmplant.krg<-krig(gsenmplant.lst,gsenmplant.
gau,15,154,se=T)
>persp(x=gsenmplant.krg$x, y=gsenmplant.krg$y,z=gsenmplant.
krg$error,
 box=F)
> title(«Estimated Standard Errors of Plant Species
Richness, n=15»)
> persp(gsenmplant.krg,box=F)
> title(«Kriged Surface for Plant Species Richness, n=15»)
```

## Spatial Autoregressive (SAR)

In analysis of spatial data, numerous attempts were made to explicitly include a spatial component into prediction models. In models with spatially correlated residuals and with autoregressive disturbance (Florax and Folmer 1992; Kelly and Gilley 1998), modeling consists of two steps. First, the response

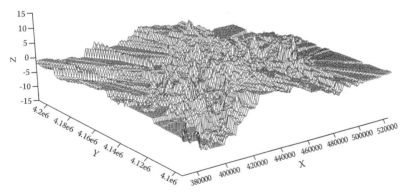

**FIGURE 4.4**
Kriged surface (3-D diagram) for plant species richness at GSENM, Utah.

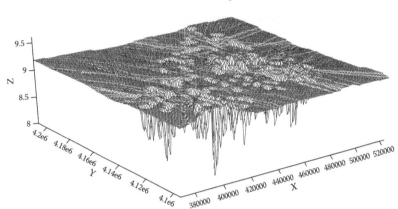

**Estimated Standard Errors of Plant Species Richness, n=15**

**FIGURE 4.5**
Standard errors (3-D diagram) for plant species richness at GSENM, Utah.

variable is treated as nonspatial and a linear model is applied. Then, the residuals of a linear model on training data are assumed to be spatially correlated and their dependence is modeled through a "neighborhood" matrix using an autoregressive approach.

Semivariograms, which describe how the sample variance changes as a function of distance, will be used to evaluate spatial dependencies among the residuals from the various models. If the residuals exhibited spatial dependencies, a spatial autoregressive (SAR) model will be used to obtain generalized least squares (GLS) estimates of the regression coefficients associated with the TS model (Upton and Fingleton 1985). The model residuals will be reevaluated to ensure the removal of the spatial dependencies. In fitting the SAR models, a spatial weight matrix (i.e., a block diagonal matrix) based on inverse distance weighting will be used to represent the spatial dependencies among the sampling units of the landscape.

## Statistical Example

```
gsenmplant[1,] List all variables (read first row)
CtTot     CvNtv     CtExo     CvTot     CtNtv     CvExo     band1
25        10.1      2         10.6      22        0.45      101
band2     band3     band4     band5     band61    band6h
96        113       71        157       163       208
band7     band8     tasl1     tasl2     tasl3     tasl4     tasl5
121       71        248       -68       -57       20        -37
tasl6     elv       slp       absasp    landform  xutm      utm
-20       1599      29.9      178.86    -0.30476  472457    4132051
```

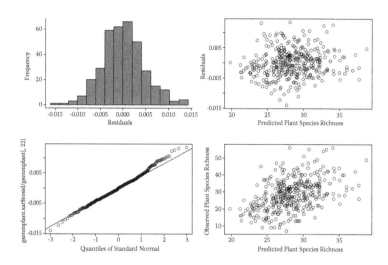

**FIGURE 4.6**

Histogram of the residuals, a scatter plot of the residuals over predicted plant species richness at GSENM area, a Q-Q plot of the residuals, and a scatter plot of observed plant species richness at GSENM area over the predicted.

Recall the OLS Model:

```
gsenmplant.ols<-ols(gsenmplant[,1],gsenmplant[,c(13,18,19)],w=
gsenmplant.wt)
```

```
Residual Standard Error = 8.7624, Multiple R-Square = 0.1109
N = 359, F-statistic = 14.7642 on 3 and 355 df, p-value = 0
```

|           | coef     | std.err | t.stat   | p.value |
|-----------|----------|---------|----------|---------|
| Intercept | 71.8834  | 6.9618  | 10.3253  | 0.0000  |
| band6h    | -0.1670  | 0.0314  | -5.3177  | 0.0000  |
| tasl3     | 0.0674   | 0.0326  | 2.0639   | 0.0398  |
| tasl4     | -0.1417  | 0.0583  | -2.4323  | 0.0155  |

```
Log(like)      = -1286.5859
AIC            = 2581.1719
AICC           = 2581.2849
Schwartz       = 2596.7052
```

```
Moran's I(res)   = 0.1183
Mean of I        = -0.1968
Std Dev of I     = 0.1906
Z-value of I     = 1.6535
P-value(2-side)  = 0.0982
```

```
Lagrange Mult      = 34.6326
P-value(2-side)    = 0
```

## Using Spatial AR Model (without Regression)

```
gsenmplant.sarwo<-spatar1(gsenmplant.ols$resid, w=gsenmplant.
wt,xreg=F)
[1] 1.126859
```

```
Residual Standard Error = 8.2469 , Multiple R-Square = 0.1042
N = 359 , F-Statistic = 20.8853 on 2 and 357 df, p-value = 0
```

|           | coef    | std.err | t.stat   | p.value |
|-----------|---------|---------|----------|---------|
| Intercept | -0.0187 | 1.1269  | -0.0166  | 0.9868  |
| lambda    | 0.6137  | 0.0972  | 6.3128   | 0.0000  |
| Variance  | 68.0107 | 5.1328  | 13.2501  | NA      |

```
Log(like)    = -1272.4216
AIC          = 2548.8432
AICC         = 2548.8769
Schwartz     = 2556.6098
```

|                              | Value    | P-Value |
|------------------------------|----------|---------|
| Likelihood Ratio Test (df=1) | 28.3287  | 0       |

## Using Spatial AR Model (with Regression, OLS Model) Using R or S-Plus

```
gsenmplant.sar<-spatar1(gsenmplant[,1],gsenmplant[,c(13,18,19)
],w=gsenmplant.wt)
```

```
[1] 7.92358513 0.03478037 0.03184303 0.06598958
```

```
Residual Standard Error = 8.204 , Multiple R-Square = 0.2118
 N = 359 , F-Statistic = 19.2982 on 5 and 354 df, p-value = 0
```

|           | coef     | std.err | t.stat   | p.value |
|-----------|----------|---------|----------|---------|
| Intercept | 61.0995  | 7.9236  | 7.7111   | 0.0000  |
| band6h    | -0.1253  | 0.0348  | -3.6022  | 0.0004  |
| tasl3     | 0.0513   | 0.0318  | 1.6125   | 0.1077  |
| tasl4     | -0.1177  | 0.0660  | -1.7835  | 0.0754  |
| lambda    | 0.6486   | 0.0925  | 7.0155   | 0.0000  |
| Variance  | 67.3064  | 5.0842  | 13.2385  | NA      |

```
Log(like)    = -1271.3053
AIC          = 2552.6105
AICC         = 2552.7805
Schwartz     = 2572.0271
```

```
                                 Value      P-Value
Likelihood Ratio Test (df=1)     30.5614    0

gsenmplant.sar<-spatar1(gsenmplant[,1],gsenmplant[,c(13,19)],w
=gsenmplant.wt)

[1]  7.94959572 0.03390253 0.05588163

Residual Standard Error = 8.2276 , Multiple R-Square = 0.2073
```

N = 359 , F-Statistic = 23.4718 on 4 and 355 df, p-value = 0

```
               coef        std.err     t.stat      p.value
Intercept      59.8432     7.9496      7.5278      0.0000
band6h         -0.1383     0.0339      -4.0787     0.0001
tasl4          -0.0601     0.0559      -1.0748     0.2832
lambda         0.6600      0.0908      7.2677      0.0000
Variance       67.6934     5.1148      13.2349     NA

Log(like)  = -1272.5943
AIC        = 2553.1885
AICC       = 2553.3015
Schwartz   = 2568.7218

                                 Value      P-Value
Likelihood Ratio Test (df=1)     32.2655    0
```

### Example on How to Develop Plot of Standard Normal Distribution

```
par(mfrow=c(2,2))
rn<-rnorm(261)
qqnorm(rn)
qqline(rn)
rn<-rnorm(261)
qqnorm(rn)
qqline(rn)
rn<-rnorm(261)
qqnorm(rn)
qqline(rn)
```

### Analysis of Residuals for Plant Species Richness (gsenmplant) Data

```
gsenmplant.pred<-gsenmplant[,1]-gsenmplant.sar$resid
par(mfrow=c(2,2))
hist(gsenmplant.sar$resid,ylab="Frequency",xlab="Residuals")
plot((gsenmplant[,1]-gsenmplant.sar$resid),gsenmplant.sar$resi
d,ylab="Residuals",xlab="Predicted Plant Species Richness")
```

```
qqnorm(gsenmplant.sar$resid)
qqline(gsenmplant.sar$resid)

plot(gsenmplant.pred,gsenmplant[,1],ylab="Observed Plant
Species Richness", xlab="Predicted Plant Species Richness")

par(mfrow=c(1,1))
plot(gsenmplant[,23],gsenmplant.sar$resid) - Observed vs.
Residuals
```

### Weighted SAR Model

```
gsenmplant.sar<-spatar1(gsenmplant[,1],cbind(1/gsenmplant[,22]
,gsenmplant[c,(19)]),w=gsenmplant.wt)
gsenmplant.pred<-gsenmplant[,1]-gsenmplant.sar$resid

par(mfrow=c(2,2))
hist(gsenmplnat,1]- gsenmplant.sar$resid),gsenmplant.
sar$resid/gsenmplant[,22],ylab="Residuals",xlab="Predicted
Plant Species Richness")

qqnorm(gsenmplant.sar$resid/gsenmplant[,22])
qqline(gsenmplant.sar$resid/gsenmplant[,22])
plot(gsenmplant.pred,gsenmplant[,1],ylab="Observed Plant
Species Richness",xlab="Predicted Plant Species Richness")
```

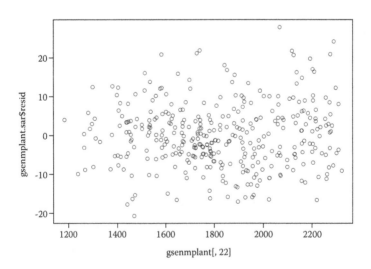

**FIGURE 4.7**
Plotting observed versus residuals for plant species richness within GSENM, Utah.

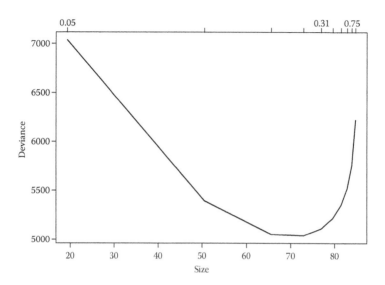

**FIGURE 4.8**
Estimated deviance to determine number of tree nodes for binary classification trees of vegetation cover types within GSENM, Utah.

## Binary Classification Tree (BCTs)

In addition to the use of kriging for modeling fine and coarse scale of spatial variability of particular landscape, regression binary classification trees (BCTs) may be used as part of spatial statistical modeling using the ordinary least squares, generalized linear model, or other statistical procedures. Regression classification tree-based models are nonparametric models that provide an alternative to linear or additive models for regression problems and to linear or additive logistic models for classification problems (MathSoft Inc. 2000). They are fit by successively splitting the data to form homogeneous subsets. The result is a hierarchical tree of decision rules useful for prediction or classification. In order to reduce the error associated with large trees (i.e., deviance), it is necessary to "prune" the tree to a simplified version where the variance is stable. By plotting deviance versus tree nodes, the optimal number of nodes with minimal deviance will be determined. These classification trees have minimal misclassification error rates.

Following is an example using classification trees for predicting vegetation map classes:

```
> veg<-spinput("c:\\gsenm\\gsenmveg.txt")
> veg[1,]
```

```
vegcode     band1      band2      band3      band4      band5      band61
5           112        110        137        90         190        174
band6h      band7      band8      tasl1      tasl2
230         158        93         298        -77
tasl3       tasl4      tasl5      tasl6      elv        slp
-75         16         -55        -21        1865.075   6.787294
absasp      landform
129.8112    -0.0607422


> veg<-data.frame(veg)
> veg[,1]<-as.factor(veg[,1])
> hveg<-tree(vegcode~.,data=veg)
> summary.tree(hveg)

Classification tree:
tree(formula = vegcode ~ ., data = veg)
```

## Variables actually used in tree construction:

```
[1] "slp" "band6h" "elv" "tasl2" "band5" "tasl1" "band61"
"band8"
[9] "tasl6" "band4" "band2" "band7" "band3" "tasl4" "absasp"
"tasl3"
[17] "band1" "tasl5"
Number of terminal nodes: 85
Residual mean deviance: 1.11 = 3268 / 2945
Misclassification error rate: 0.2215 = 671 / 3030


> hveg.pred<-predict.tree(hveg,newdata=veg,type="class")
> hveg.cv<-cv.tree(hveg)
> hveg.cv

$size:
 [1] 19.37170 50.52975 65.61413 72.83067 76.91989 79.55778
81.40226 82.76517
 [9] 83.81365 84.64540

$dev:
 [1] 7036.468 5394.613 5052.521 5039.302 5110.632 5217.775
5351.884 5520.808
 [9] 5756.970 6219.911

$k:
 [1] 0.0500000 0.1052632 0.1666667 0.2352941 0.3125000
0.4000000 0.5000000
 [8] 0.6153846 0.7500000 0.9090909

attr(, "class"):
[1] "shrink" "tree.sequence"
```

```
> summary(hveg.cv)

 Length Class Mode
size 10 numeric
 dev 10 numeric
  k 10 numeric

> plot(hveg.cv)
```

Develop the error matrix table:

```
>table(veg[,1],hveg.pred)
```

|   | 1 | 2 | 3 | 4 | 5 | 6 | 7 | 8 |
|---|---|---|---|---|---|---|---|---|
| 1 | 474 | 21 | 105 | 1 | 21 | 0 | 1 | 17 |
| 2 | 42 | 330 | 66 | 3 | 10 | 0 | 2 | 7 |
| 3 | 76 | 93 | 1279 | 0 | 12 | 2 | 3 | 15 |
| 4 | 12 | 0 | 9 | 18 | 0 | 0 | 0 | 1 |
| 5 | 13 | 1 | 21 | 0 | 91 | 0 | 0 | 4 |
| 6 | 10 | 1 | 16 | 0 | 0 | 26 | 6 | 1 |
| 7 | 3 | 0 | 7 | 0 | 0 | 0 | 20 | 0 |
| 8 | 21 | 9 | 27 | 0 | 6 | 8 | 0 | 119 |

The next step is to prune the tree based on the lowest value of deviance (small error) and this can be done as:

```
> hveg.prune<-prune.tree(hveg,best=72)
> plot(hveg.prune)
> text(hveg.prune,cex=0.5)
> summary.tree(hveg.prune)

Classification tree:
snip.tree(tree = hveg, nodes = c(440, 636, 637, 148, 96, 1012,
28, 1090))
```

Variables actually used in tree construction:

```
[1] "slp" "band6h" "elv" "tasl2" "band5" "tasl1" "band6l"
"tasl6"
[9] "band4" "band2" "tasl4" "absasp" "tasl3" "band1" "tasl5"
Number of terminal nodes: 72
Residual mean deviance: 1.227 = 3630 / 2958
Misclassification error rate: 0.2416 = 732 / 3030

> hveg.prune (Partial output for Binary Classification Trees)
node), split, n, deviance, yval, (yprob)
 * denotes terminal node

 1) root 3030 8810.000 3 ( 0.21120 0.15180 0.48840 0.013200
0.042900 0.019800 0
.009901 0.062710 )
```

```
 2) slp<6.32762 1751 5137.000 3 ( 0.26560 0.16790 0.42150
0.016560 0.024560 0
.018280 0.001713 0.083950 )
 4) band6h<244.5 1490 4459.000 3 ( 0.27250 0.18660 0.38320
0.007383 0.02886
0 0.021480 0.002013 0.097990 )
 8) elv<2039.3 1043 3075.000 3 ( 0.32410 0.14290 0.36720
0.010550 0.03164
0 0.009588 0.002876 0.111200 )
 16) elv<1451.4 81 112.200 3 ( 0.00000 0.48150 0.51850
0.000000 0.00000
0.000000 0.000000 0.000000 0.000000 0.000000 ) *
......
 2031) absasp>170.791 8 0.000 2 ( 0.00000 1.00000 0.00000 0
.000000 0.000000 0.000000 0.000000 0.000000 ) *
 127) elv>2203.85 67 148.000 3 ( 0.04478 0.02985 0.62690
0.000000 0.
149300 0.000000 0.149300 0.000000 )
 254) tasl4<16.5 12 10.810 7 ( 0.00000 0.16670 0.00000
0.000000 0
.000000 0.000000 0.833300 0.000000 ) *
 255) tasl4>16.5 55 74.200 3 ( 0.05455 0.00000 0.76360
0.000000 0
.181800 0.000000 0.000000 0.000000 ) *
```

Develop the final error matrix table:

```
> hveg.pred1<-predict.tree(hveg.prune,newdata=veg,type="class")
> table(veg[,1],hveg.pred1)
```

|   | 1   | 2   | 3    | 4  | 5  | 6  | 7  | 8   |
|---|-----|-----|------|----|----|----|----|-----|
| 1 | 452 | 11  | 139  | 1  | 21 | 0  | 1  | 15  |
| 2 | 55  | 212 | 171  | 3  | 10 | 0  | 2  | 7   |
| 3 | 51  | 32  | 1368 | 0  | 12 | 2  |    | 12  |
| 4 | 12  | 0   | 9    | 18 | 0  | 0  | 0  | 1   |
| 5 | 13  | 1   | 28   | 0  | 84 | 0  | 0  | 4   |
| 6 | 11  | 0   | 16   | 0  | 0  | 26 | 6  | 1   |
| 7 | 3   | 0   | 7    | 0  | 0  | 0  | 20 | 0   |
| 8 | 21  | 9   | 25   | 0  | 6  | 8  | 0  | 121 |

## Cokriging

Cokriging allows one or more secondary variables to be included in the model and assuming that the primary and secondary variables are moderately correlated, the estimation accuracy of the primary variable should increase. Cokriging estimation ensures that the value of a variable estimated,

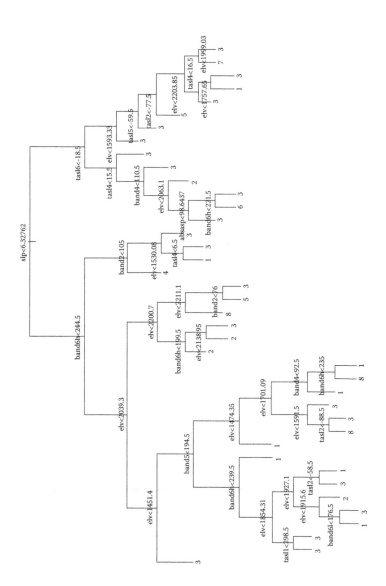

**FIGURE 4.9**
Representative output for binary classification trees for final vegetation cover types map at GSENM, Utah.

on the basis of the neighboring values of one or several other variables, is the best possible based on the following criteria:

a. The absence of bias between the estimated value and the true one

b. The minimization of the variance of the estimations

Kanevski and Maignan (2004) stated that several types of cokriging models can be distinguished as:

- Ordinary cokriging
- Standardized ordinary cokriging
- Simple cokriging

In all the types of cokriging, the sum of the weights for the main variable is equal to one, while the sum for the secondary variable is equal to zero. Sometimes an improvement can be reached by introducing a new additional variable with the same mean value as for the main variable. In this case the sum of all weights is equal to one (Olea 1999; Kanevski and Maignan 2004).

In order to develop a cokriging, we will use standard distance weighting measures for all spatial data. This refers to Inverse Distance Weighted (IDW, Equation 3.6) and enforces that the estimated value of a point is influenced more by nearby known points than those farther away. The method relies on a more sophisticated method of kriging called cokriging. This method uses one or more secondary features that are usually spatially correlated with the primary feature (e.g., climate variables, disturbance, solar radiation). "If the secondary features have more dense sample sets than the less intensively captured primary feature, with cokriging it can be estimated with higher accuracy without any surplus expenditure," according to Sárközy. As with kriging, the first step in the computation is the so-called experimental semivariogram using the following formula (Isaaks and Srivastava 1989; Sárközy 1999):

$$\gamma(h) = \frac{1}{2n(h)} \sum_{i=1}^{n(h)} [F(x_i) - F(x_i + h)]^2 \tag{4.5}$$

where $\gamma(h)$ is the estimated semivariance for the distance $h$, $n(h)$ is the number of measured point pairs in the distance class $h$, $F(.)$ is a measured value in (.). The equation is relatively easily computable if the measured points are ordered in a regular grid and the field has isotropy, that is $\gamma(h)$ depends only on $h$, but not on its direction. If the known points are located not regularly distance classes have to be formed, and in the lack of isotropy different semivariograms should be constructed in the typical groups of directions.

Also, according to Sárközy (1994, 1999), in the next step the experimental semivariogram(s) should be approximated by some kind of functions fitted to the experimental data by means of the least squares method (although other models are possible). But, the real power, according to Sárközy, is that cokriging uses semivariograms for each variable and cross-variograms for the variable couples. In the simple case of two variables, we have $n$ sample points of the primary variable and $m$ sample points of the single secondary variable (for simple writing we suppose $m = n + 1$) as follows:

$$W_O = K^{-1}V_O, \tag{4.6}$$

where $W_0^+ = \left|\lambda_{1,0}s_{1,0}\cdots\lambda_{n,0}0s_{m,0}\mu_1\mu_2\right|$ (the notation $s_{i0}$ stands for the weights of the secondary phenomenon, $\mu_1\mu_2$ are the Lagrange multipliers), $V_0^+ = \left|C_{0,1},Q_{1,0}\cdots C_{0,n},Q_{0,n}0Q_{0,m}11\right|$, and

$$k = \begin{vmatrix} c_{1,1} & 0 & c_{1,2} & cr_{1,2} & \cdots & c_{1,n} & cr_{1,n} & 0 & 0 & 1 & 0 \\ 0 & q_{1,1} & cr_{1,2} & q_{1,2} & \cdots & cr_{1,n} & q_{1,n} & 0 & q_{1,m} & 0 & 1 \\ \vdots & & & \vdots & & \vdots & & 0 & & & \vdots \\ c_{n,1} & 0 & c_{n,2} & cr_{n,2} & \cdots & c_{n,n} & cr_{n,n} & 0 & 0 & 1 & 0 \\ 0 & q_{n,1} & cr_{n,2} & q_{n,2} & \cdots & cr_{n,n} & q_{n,n} & 0 & q_{n,m} & 0 & 1 \\ 0 & 0 & 0 & 0 & \cdots & 0 & 0 & 0 & 0 & 0 & 0 \\ 0 & q_{m,1} & 0 & q_{m,2} & \cdots & 0 & q_{m,n} & 0 & q_{m,m} & 0 & 1 \\ 1 & 0 & 1 & 0 & \cdots & 1 & 0 & 0 & 0 & 0 & 0 \\ 0 & 1 & 0 & 1 & \cdots & 0 & 1 & 0 & 1 & 0 & 0 \end{vmatrix}$$

The coefficients have to be computed as follows: $ci, j = \gamma^{primary}(H) - \gamma^{primary}(h)$; the same is valid also for the coefficients in capital letters ($Q_{0j}$, $C_{0j}$) related to the interpolated point; $cr_{i,j} = \gamma^{ps}(H) - \gamma^{ps}(h)$ where $\gamma^{ps}$ is the regular representation of the empirical cross-variogram:

$$\gamma^{ps}(h) = \frac{1}{2n(h)}\sum_{i=1}^{n(h)}\left[F^{primary}(x_i) - F^{primary}(x_i + h)\right]\left[F^{secondary}(x_i) - F^{secondary}(x_i + h)\right]. \tag{4.7}$$

In the brief review by Sárközy (1999), many details are lacking. However, the predictive power beyond typical kriging is readily apparent. The primary feature's trend surface in the plain $f(\mathbf{x}) = A + Bx + Cy$, can be greatly improved by many secondary features (covarying predictive variables) on other plains. This is especially important for modeling patterns of biodiversity since many predictor variables are cross-correlated and autocorrelated.

Looking at the advantages of cokriging, there are some important points to consider:

- When we have several correlated variables, very often information on secondary and additional data can surround the main data set (Kanevski and Maignan 2004). Kanevski and Maignan (2004) stated that this may help to improve the estimation and a reduction in estimation variance near the border. This means that we move from an extrapolation problem to interpolation problem when using a secondary data set.

- Different monitoring networks can have very different spatial and dimensional resolutions. In general, information from the secondary monitoring network with better spatial and dimensional resolution can be more important for predictions than uncertainties related to the structural analysis (Olea 1999; Kanevski and Maignan 2004). This question has not yet been studied in detail.

## Geospatial Models for Presence and Absence Data

Most statistical models (i.e., empirical or geospatial) used for predicting species invasion and distribution (exotic biotic) focus on ecosystem characteristics to anticipate habitat suitability (species occurrence data, i.e., presence-absence versus presence-only or abundance) on the landscape (Phillips et al. 2004; Elith et al. 2006); while fewer approaches integrate species traits to gauge a species' potential invasiveness (Sakai et al. 2001). Identifying habitat suitability relies on the ability to define ecosystem characteristics (i.e., elevation, distance to water source, remote sensing spectral reflectance, and plant species richness) that may be conducive or prohibitive to invasion. The technological advancements in computer technology (i.e., memory, speed, storage, other), sound GIS, high spatial resolution of the satellite imagery data, and increased availability of geospatial data have facilitated the development of habitat suitability models that are driven by species occurrence information. Some models are fit with presence-only data (i.e., Maxent, GARP; Elith et al. 2006; Phillips et al. 2006), whereas others rely on presence and absence data (i.e., logistic regression, regression tree; Pearce and Ferrier 2000; Morissette et al. 2006). Models that are fit with presence-only data have the advantage of integrating a wide array of data that may have been collected on an opportunistic basis, archived in historical records, or limited in sample size (Ponder et al. 2001; Phillips et al. 2004; Pearson et al. 2007). Models that are fit with presence and absence data are viewed by some researchers as having greater predictive power because of their ability to detect ecological contrasts that may highlight barriers of invasion (Brotons et al. 2004; Morissette et al. 2006). Model selection for invasive species can be challenging and may often be determined by data availability and collection methods. The utility, performance, and analytical design of presence-only and presence/absence models are regularly debated in the literature (Guisan et al. 1999; Brotons et al. 2004; Guisan and Thuiller 2005; Elith et al. 2006; Pearson et al. 2007; Phillips et al. 2006; Guisan et al. 2007a, 2007b) further complicating the decision-making

**TABLE 4.1**

A Summary of the Most Widely Used Models in Ecology Ecosystems Studies and the Associated Advantages and Disadvantages of Each

| Model | Citation | Advantages | Disadvantages |
|---|---|---|---|
| GARP | Stockwell and Noble 1992; Stockwell and Peters 1999 | Presence only, nonlinear, nonparametric, not sensitive to multicollinearity, widely used | Presence only, time consuming, no variable importance analysis |
| Maxent | Phillips et al. 2006 | Presence only, nonlinear, nonparametric, not sensitive to multicollinearity, provides variables' relative importance (jackknifing), easy to run and takes less time, becoming popular | Presence only (no consideration of absence data) |
| Classification and regression tree (CART) | Breiman et al. 1984 | Nonparametric, presence/ absence, easy to run and interpret | Absence data needed |
| Logistic regression | McCullagh and Nedler 1989 | Widely used, presence/ absence | Absence data needed, sensitive to multicollinearity |
| Envelope | Jarnevich et al. 2007 | Presence-only or absence-only models can be run | All environmental factors are given equal weighting |

process. One can evaluate the most widely used models in ecology (Table 4.1) to determine how they respond to different species and compare against one another. Brief descriptions of the most common models are described in the following.

## GARP Model

GARP is the genetic algorithm for rule-set prediction model (Desktop GARP; http://nhm.ku.edu/desktopgarp/) identifies relationships between known occurrences of a species to environmental conditions or parameters (Stockwell and Nobel 1992). GARP modeling is founded on a machine-learning approach using a suite of rules (or if–then relationships) based on preconditions; if the conditions are not met, then the rules are not used. Each rule implements different approaches (i.e., atomic, envelope, negated envelope, and logistic regression) for predicting species occurrence (see Payne and Stockwell 2006). The set of rules used for the modeling process is determined using a genetic algorithm that refines the model in an evolutionary process that tests the rules on random subsets of the data. The genetic algorithm may run as many as 1000 iterations modifying the rules at random until the model's fitness no longer improves. Once the iterations are completed, the user combines the best subset of models, with predicted areas closest to the median, to generate the final surface (Anderson et al. 2003; Pearson et al. 2007). Although GARP requires presence-only data by the user, the software generates 1250 pseudo-absence points from background pixels within the study area to facilitate the genetic algorithm and for evaluating model performance. GARP evaluates performance by two processes: the correct classification rate and a confusion matrix (Fielding and Bell 1997; Anderson et al. 2003). The predictive contributions of the independent variables are not provided by the software; however, jackknife or other analyses may be performed independently by the user.

## Maxent Model

The maximum entropy distribution model (Maxent; version 3.1 or update version) was designed as a general-purpose predictive model that can be applied to incomplete and small data sets (Phillips et al. 2004, 2006). Maxent model is freely distributed on the web (http://www.cs.princeton. edu/~schapire/maxent/). The model is a maximum-entropy-based machine learning program that estimates the probability distribution for a species' occurrence based on environmental constraints (Phillips et al. 2006). It requires only species presence data (not absence) and environmental variable (continuous or categorical) layers for the study area. In addition, this model can be used for making inferences from available data while avoiding unfounded constraints from the unknown (Phillips et al. 2006). Shannon (1948) describes entropy as the measure of uncertainty associated with a random variable; thus the greater the entropy, the lower the amount of known information. Adhering to these concepts, Maxent utilizes presence-only points of occurrence avoiding absence data and evading assumptions on the range of a given species (plants or animals). Predictions are presented in either 0:1 for continuous raw output, or 0:100 for continuous cumulative output. Evaluation of the Maxent model is performed in several ways by the program. First, the user has the option of defining a percentage of the data for validation that (1) plots testing and training omissions against thresholds; (2) plots predicted area against threshold; and (3) generates the receiver operating characteristic curve (ROC) for area under the curve (AUC) evaluation. Second, a jackknife option allows the estimation of the bias and standard error in the statistics, and tests the predictive contribution of independent variables. To conclude, Maxent generates response curves that show individual relationships between each independent variable and the model's prediction.

## Logistic Regression

Binomial logistic regression analyses are commonly used by ecologists for predictive modeling (Austin et al. 1990; Pearce and Ferrier 2000). Based on Bernoulli's discrete probability distribution (Evans et al. 2000), logistic regression analysis is a generalized linear model that uses the logit (the natural log of the odds of the dependent occurring or not) as its link function (McCullagh and Nelder 1989). Requiring presence and absence data, logistic regression analysis is used to predict a dependent variable from a number of continuous or categorical independent variables ranking their relative importance and assessing the degrees of interaction. The dependent variable is usually dichotomous taking a value of 1 with a probability of success ($p$) and value of 0 with a probability of failure ($1 - p$). This method also applies a maximum likelihood estimation that quantifies the probability of a certain event occurring; in this case, the probability of occurrence by an invasive species. The performance of a logistic regression model is generally assessed by a D-squared value and goodness-of-fit tests (i.e., chi-square and Wald statistics).

Predicted values are transformed using the following formula:

$$\text{Probability} = \text{Exp (linear predictor)}/1 + \text{Exp (linear predictor)} \quad (4.8)$$

where *Probability* ranges from 0 to 1, *Exp* is a term refers as exponential, and the *linear predictors* are the significant independent variables used in the model. Standardized regression coefficients are used to rank the predictive influence of each independent variable, while the raw model coefficients can be used to create a modeled surface.

## Classification and Regression Tree (CART)

Classification and regression tree analysis is a commonly used nonparametric model that predicts the response of a dependent variable through a series of simple regression analyses (Breiman et al. 1984). Unlike other regression approaches that conduct simultaneous analyses, classification trees statistically partition the dependent data into two homogenous groups at a node, repeating the procedure for each group in a continuing process that forms R Classification trees that are used when the dependent data is categorical (i.e., presence, absence), and function similarly to regression trees which use continuous dependent data (i.e., percent cover, species richness). There are several characteristics of this modeling approach that are appealing to researchers and resource managers. First, the analyses explicitly allow for nonlinear relationships between the dependent and independent variable. These methods make no a priori assumptions about the distribution of the data, the relationships among independent variables, or relationships between the dependent and independent variables. Second, they are well suited to handle nonhomogeneous data sets (i.e., unbalanced sample sizes, high variability). Finally, the results are easily interpreted and the predictive strength of each independent variable is explicitly reported.

## Envelope Model

Envelope is a simple modeling tool that is supported by ArcGIS 9x (ESRI 2007) and available on the National Institute of Invasive Species Science Web site (see Jarnevich et al. 2007; http://www.niiss.org). Originally designed to increase field sampling efficiency, Envelope's output does not rely on any statistical applications but simply defines the parameters of the environmental independent variables for the survey data. For each presence point, the values of each independent variable is extracted, with the minimum and maximum values of all presence points defining the probability range of occurrence for each variable. For example, if the presence data for a species falls between elevations of 1800 m and 2200 m, then all pixels that are within that elevation range will define the area where a species may occur, whereas pixels with values <1800 m or >2200 m would fall outside the probability range. This process is repeated with each candidate variable and the results summed in a final output surface. The output surface is in a grid format with each pixel ranked according to the number of variables that are within the probability range. All candidate variables are assumed to be predictors and are weighted equally in the model's output. Envelope lacks the complexity and statistical applications used by other models to identify the potential significance and predictive contribution of each variable; however, surfaces are generated for each independent variable and can be evaluated individually for pixels that fall within or outside of the range.

---

## References

Agterburg, F. P. 1984. Trend surface analysis. In *Spatial Statistics and Models*, G. L. Gaile and C. J. Willmott (eds.), 147–171. Dordrecht, The Netherlands: Reidel.

Akaike, H. 1997. "On entropy maximization principle," In *Applications of Statistics*, P. R. Krishnaiah (ed.), 27–41. Amsterdam, NorthHolland.

Anderson, R. P., D. Lew, and A. T. Peterson. 2003. Evaluating predictive models of species' distributions: Criteria for selecting optimal models. *Ecological Modeling* 162:211–232.

Armstrong, M, 1992. Freedom of speech? *De Geostatisticis* July, no. 14.

Armstrong, M., and N. Champigny. 1988. A study on kriging small blocks. *CIM Bulletin* 82:923.

Austin, M. P., A. O. Nicholls, and C. R. Margules. 1990. Measurement of the realized qualitative niche: Environmental niche of five Eucalyptus species. *Ecological Monograms* 60:161–177.

Breiman, L., J. H. Friedman, R. A. Olshen, and C. G. Stone. 1984. *Classification and Regression Trees*. Belmont, CA: Wadsworth International Group.

Brockwell, P. J., and R. A. Davis. 1991. *Time Series: Theory and Methods*. New York: Springer.

Brotons, L., W. Thuiller, M. B. Araujo, and A. H. Hirzel. 2004. Presence-absence versus presence-only modeling methods for predicting bird habitat suitability. *Ecography* 27:437–448.

Champigny, N. 1992. Geostatistics: A tool that works. *The Northern Miner*, May 18.

Chen, S., C. F. N. Cowan, and P. M. Grant. 1991. Orthogonal least squares learning for radial basis function networks. *IEEE Transactions on Neural Networks* 2:302–309.

Cliff, A., and J. K. Ord. 1981. *Spatial Processes, Models and Applications*. London: Pio, Ltd.

Cressie, N. A. C. 1993. *Statistics for Spatial Data*. New York: Wiley-Interscience.

David, M, 1977. *Geostatistical Ore Reserve Estimation*. Elsevier Scientific Publishing Company, Amsterdam.

De'Ath, G., and K. E. Fabricus. 2000. Classification and regression trees: A powerful yet simple technique for ecological data analysis. *Ecology* 81:3178–3192.

Derksen, S., and H. J. Keselman. 1992. Backward, forward and stepwise automated subset selection algorithms: Frequency of obtaining authentic and noise variables. *British Journal of Mathematical and Statistical Psychology* 45:265–282.

Deutsch, C.V., and Journel, A. G. 1992. *GSLIB - Geostatistical Software Library and User's Guide*. Oxford University Press, New York.

Dobrowski, S. Z., J. A. Greenberg, C. M. Ramirez, and S. L. Ustin. 2005. Improving image-derived vegetation maps with regression-based distribution modeling. *Ecological Modeling* 192:126–142.

Draper, N. R., and H. Smith. 1981. *Applied Regression Analysis*, 2nd ed. New York: Wiley.

Efron, B., and R. J. Tibshirani. 1993. *An Introduction to the Bootstrap*. New York: Chapman & Hall.

Elith, J., C. J. Graham, R. Anderson, M. Dudik, S. Ferrier, A. Guisan, R. Hijmans, F. Huettmann, J. Leathwick, A. Lehmann, J. Li, L. Lohmann, B. Loiselle, G. Manion, C. Moritz, M. Nakamura, Y. Nakazawa, J. Overton, A. T. Peterson, S. Phillips, K. Richardson, R. Scachetti-Pereia, R. Schapire, J. Soberon, S. Williams, M. Wisz, and N. Zimmermann. 2006. Novel methods improve prediction of species distributions from occurrence data. *Ecography* 29:129–151.

ESRI. 1995. *ARC/INFO® Software and On-Line Help Manual*. Redlands, CA: Environmental Research Institute.

ESRI. 2007. *ArcGIS 9.x*. ESRI, Redlands, CA. http://www.esri.com/index.html

Evans, M., N. Hastings, and B. Peacock. 2000. Bernoulli distribution. In *Statistical Distributions*, 3rd ed., 31–33. New York: Wiley.

Florax, R., and H. Folmer. 1992. Specification and estimation of spatial linear regression models—Monte Carlo evaluation of pre-test parameters. *Regional Science and Urban Economics* 22:405–432.

Fortin, Marie-Josee, and M. R. T. Dale. 2005. *Spatial Analysis: A Guide for Ecologists*. Cambridge: Cambridge University Press.

Frank, I. E., and J. H. Friedman. 1993. A statistical view of some chemometrics regression tools. *Technometrics* 35:109–148.

Freedman, L. S., D. Pee, and D. N. Midthune. 1992. The problem of underestimating the residual error variance in forward stepwise regression. *The Statistician* 41:405–412.

Friedl, M. A., and C. E. Brodley. 1997. Decision tree classification of land cover from remotely sensed data. *Remote Sensing and the Environment* 61:399–409.

Furnival, G. M., and R. W. Wilson. 1974. Regression by leaps and bounds. *Technometrics* 16:499–511.

Gentleman, W. M. 1974. Basic procedures for large sparse or weighted least-squares. *Applied Statistics* 23:448–454.

Goodnight, J. H. 1979. A tutorial on the SWEEP Operator. *The American Statistician* 33:149–158.

Goovaerts, P. 1997. *Geostatistics for Natural Resources Evaluation* (Applied Geostatistics Series). New York: Oxford University Press.

Gown, S. N., R. H. Waring, D. G. Dye, and J. Yang. 1994. Ecological remote sensing at OTTER: Satellite macroscale observation. *Ecological Applications* 4:322–343.

Guisan, A., C. H. Graham, J. Elith, F. Huettmann, and the NCEAS Species Distribution Modelling Group. 2007a. Sensitivity of predictive species distribution models to change in grain size. *Diversity and Distributions* 13:332–340.

Guisan A., and W. Thuiller. 2005. Predicting species distribution: Offering more than simple habitat models. *Ecology Letters* 8(9):993–1009.

Guisan, A., S. B. Weiss, and A. D. Weiss. 1999. GLM versus CCA spatial modeling of plant species distribution. *Plant Ecology* 143:107–122.

Guisan, A., and N. E. Zimmermann. 2000. Predictive habitat distribution models in ecology. *Ecological Modelling* 135:47–186.

Guisan, A., N. E. Zimmermann, J. Elith, C. H. Graham, S. Phillips, and A. T. Peterson 2007b. What matters for predicting the occurrences of trees: Techniques, data or species' characteristics? *Ecological Monograms* 77:615–630.

Hald, A. 1952. *Statistical Theory with Engineering Applications.* New York: John Wiley & Sons.

Hansen, M., R. Dubayah, and R. de Fries. 1996. Classification trees: An alternative to traditional land cover classifiers. *International Journal of Remote Sensing* 17(5):1075–1081.

Hartigan, J. A. 1975. *Clustering Algorithms.* New York: John Wiley & Sons.

Hevesi, J. A., J. D. Istok, and A. L. Flint. 1992. Precipitation estimation in mountainous terrain using multivariate geostatistics. Part I: Structural analysis. *Journal of Applied Meteorology* 31:661–676.

Holmgren, P., and T. Thuresson. 1998. Satellite remote sensing for forestry planning: A review. *Scandinavian Journal of Forest Research* 13:90–110.

Isaaks, E. H., and R. M. Srivastava. 1989. *An Introduction to Applied Geostatistics.* New York: Oxford University Press.

Jarnevich, C. S., J. Graham, G. Newman, A. Crall, and T. J. Stohlgren. 2007. Balancing data sharing requirements for analyses with data sensitivity. *Biological Invasions* 9:597–599.

Kalkhan, M. A., E. J. Stafford, and T. J. Stohlgren. 2007b. Rapid plant diversity assessment using a pixel nested plot design: A case study in Beaver Meadows, Rocky Mountain National Park, Colorado, USA. *Diversity and Distributions* 13:379–388.

Kalkhan, M. A., E. J. Stafford, P. J. Woodly, and T. J. Stohlgren. 2007a. Exotic plant species invasion and associated abiotic variables in Rocky Mountain National Park, Colorado, USA. *Journal of Applied Soil Ecology* 37:622–634.

Kalkhan, M. A., G. W. Chong, R. M. Reich, and T. J. Stohlgren. 2000. Landscape-scale assessment of mountain plant diversity: Integration of Remotely Sensed Data, GIS, and Spatial Statistics. In: ASPRS Annual Convention & Exposition, ASPRS Technical Papers, May 22–26, 2000, Washington, DC. ASPRS- CD -ROMs Publication (ISBN 1-57083-061-4) by Clearance Center, Inc, 222 Rosewood Drive, Danvers, MA 01923 (Adobe Acrobat Reader Format, p 163).

Kalkhan, M. A., E. J. Martinson, P. N. Omi, T. J. Stohlgren, G. W. Chong, and M. A. Hunter. 2004. Integration of spatial information and spatial statistics: a case study of invasive plants and wildfire on the Cerro Grande fire, Los Alamos, New Mexico. Pages 191–199. In R. T. Engstrom, K. E. M. Galley, and W. J. de Groot (eds.). Proceedings of the 22nd Tall Timbers Fire Ecology Conference: Fire in Temperate, Boreal, and Montane Ecosystems. Tall Timbers Research Station, Tallahassee, FL.

Kalkhan, M. A., T. J. Stohlgren, G. W. Chong, Lisa D. Schell, and R. M. Reich. 2001. A predictive spatial model of plant diversity: Integration of Remotely Sensed data, GIS, and Spatial statistics. Proceeding of the Eight Forest Remote Sensing Application Conference (RS 2000), April 10–14, 2000, Albuquerque, New Mexico, 11 pp. CD-ROMs Publications (ISBN 1-57083-062-2).

Kallas, M. 1997. Hazard rating of Armillaria root rot on the Black Hills National Forest. M.S. Thesis, Department of Forest Sciences, Colorado State University, Fort Collins, CO.

Kanevski, M., and M. Maignan. 2004. *Analysis and Modeling of Spatial Environmental Data.* New York: Marcel Dekker.

Kelly, R. P., and O. Gilley. 1998. Generalizing the OLS and grid estimators. *Real Estate Economics* 26:331–347.

Kravchenko, A., and D. G. Bullock. 1999. A comparative study of interpolation methods for mapping soil properties. *Agronomy Journal* 91:393–400.

Lawson, C. L., and Hanson, R. J. 1974. *Solving Least Squares Problems.* Englewood Cliffs, NJ: Prentice-Hall. (2nd ed., Philadelphia: SIAM, 1995.)

Lipschutz, S. 1968. *Theory and Problems of Probability.* New York: McGraw-Hill.

Mackenthum, K. M. 1976. *A Statement on Phosphorus.* Washington, DC Office of Water Planning and Standards, USEPA.

Matheron, G. 1962. *Traité de géostatistique appliquée. Tome 1.* Editions Technip, Paris.

MathSoft Inc. 2000. *Insightful S-Plus.* Seattle, WA: MathSoft Inc.

Maxent Model. 2007. Version 3.1; http://www.cs.princeton.edu/~schapire/maxent

McCullagh, P., and J. A. Nelder. 1989. *Generalized Linear Models,* 2nd ed. New York: Chapman & Hall.

Metzger, K., 1997. Modeling small-scale spatial variability in stand structure using remote sensing and field data. M.S. Thesis, Department of Forest Sciences, Colorado State University, Fort Collins, CO.

Miller, A. J. 1984. Selection of subsets of regression variables (with discussion). *Journal of the Royal Statistical Society, Series A* 147:389–425.

Miller, A. J. 1990. *Subset Selection in Regression.* London: Chapman & Hall.

Morissette, J. T., C. S. Jarnevich, A. Ullah, W. Cai, J. A Pedelty, J. E. Gentle, T. J. Stohlgren, and J. L. Schnase. 2006. A tamarisk habitat suitability map for the continental United States. *Frontiers in Ecology and the Environment* 4:11–17.

Myers, R. H. 1986. *Classical and Modern Regression with Applications.* Boston: Duxbury Press.

Olea, R. A. 1999. *Geostatistics for Engineers and Earth Scientists.* Norwell, MA: Kluwer Academic Publishers.

Orr, M. J. L. 1995. Regularisation in the selection of radial basis function centres. *Neural Computation* 7:606–623.

Orr, M. J. L. 1996. Introduction to radial basis function networks. http://www.cns.ed.ac.uk/people/mark/intro.ps; http://www.cns.ed.ac.uk/people/mark/intro/intro.html.

Osborne, M. R. 1976. On the computation of stepwise regressions. *Australia Computer Journal* 8:61–68.

Payne, K., and D. R. B. Stockwell. 2006. *GARP Modeling System User's Guide and Technical Reference.* http://www.landshape.org/enm/

Pearce, J., and S. Ferrier. 2000. Evaluating the predictive performance of habitat models developed using logistic regression. *Ecological Modeling* 133:225–245.

Pearson, R. G., C. J. Raxworthy, M. Nakamura, and A. T. Peterson. 2007. Predicting species distribution from small numbers of occurrence records: A test case using cryptic geckos in Madagascar. *Journal of Biogeography* 34:102–117.

Philip, G. M., and D. F. Watson. 1986. Matheronian statistics—Quo vadis? *Mathematical Geology* 18(1): 93-117

Phillips, S. J., R. P. Anderson, and R. E. Schapire. 2006. Maximum entropy modeling of species geographic distributions. *Ecological Modeling* 190:231–259.

Phillips, S. J., M. Dudik, and R. E. Schapire. 2004. A maximum entropy approach to species distribution modeling. In *Proceedings of the 21st International Conference on Machine Learning,* 655–662. New York: ACM Press.

Ponder, W. F., G. A. Carter, P. Flemons, and R. R. Chapman. 2001. Evaluation of museum collection data for use in biodiversity assessment. *Conservation Biology* 15:648–657.

Reich, R. M., and R. A. Davis. 1998. *On-Line Spatial Library for the S-Plus Statistical Software Package.* Colorado State University: Fort Collins, CO. (Available at: http://www.warnercnr.colostate.edu/~robin/)

Reich, R. M., C. Aguirre-Bravo, M. A. Kalkhan, and V. A. Bravo. 1999. Spatially based forest inventory for Ejido El Largo, Chihuahua, Mexico. In: North America Science Symposium: Toward a unified framework for inventorying and monitoring forest ecosystem resources, November 1–6, 1998, Guadalajara, Jalisco, Mexico, USDA Forest Service, Rocky Mountain Research Station, Proceedings RMRS-P-12, December 1999, p. 31–41.

Robertson, G. P. 1987. Geostatistics in ecology: Interpolating with known variance," *Ecology*, Vol. 63,pp. 744–748.

Roecker, E. B. 1991. Prediction error and its estimation for subset-selected models. *Technometrics* 33:459–468.

Sakai, A. K., F. W. Allendorf, J. S. Holt, D. M. Lodge, J. Molofsky, K. A. With, S. Baughman, et al. 2001. The population biology of invasive species. *Annual Review of Ecology and Systematics* 32:305–332.

Sárközy F. 1994. The GIS concept and the 3-dimensional modeling. *Computers, Environment and Urban Systems* 18(2):111–121.

Sárközy, F., and P. Gáspár. 1992. Modelling of scalar fields represented by scattered 3D points. *Periodica Polytechnica Civil Engineering* 36(2):187–200.

Sárközy, F., and J. Závoti. 1995 Conceptional data model for modeling of scalar fields and one compression method usable for its implementation. *Proceedings of the Fourth International Symposium of LIESMARS,* China, October 25–27, pp. 1–10.

Sárközy F., and J. Závoti. 1999. GIS functions interpolations. *Periodica Polytechnica Civil Engineering* 43(1):63–86.

Sárközy F. 1999. GIS Function. *Periodica Polytechnica,* SER. CIV. ENG. VOL. 43, pp. 87–106.

Schabenberger, O., and C. A. Gotway. 2005. *Statistical Method for Spatial Data Analysis.* Boca Raton, FL: Chapman & Hall/CRC Press.

Schabenberger, O., and F. J. Pierce. 2001. *Contemporary Statistical Models for the Plant and Soil Sciences.* Boca Raton, FL: CRC Press.

Shannon, C. E. 1948. A mathematical theory of communication. *Bell System Technical Journal* 27:379–423, 623–656.

Shepard, D. 1968. A two-dimensional interpolation function for irregularly-spaced data. Proceedings of the 1968 ACM National Conference, pp. 517–524. Soares, A. 1992. Geostatistical estimation of multi-phase structures. *Mathematical Geology* 24:149–160.

Stockwell, D., and I. R. Nobel. 1992. Induction of sets of rules from animal distribution data: A robust and informative method of data analysis. *Mathematics and Computer in Simulation* 33:385–390.

Ullah, U. 1998. *Handbook of Applied Economic Statistics.* New York: Marcel Dekker.

Underwood, E. C., R. Klinger, and P. E. Moore. 2004. Predicting patterns of non-native plant invasions in Yosemite National Park, California, USA. *Diversity and Distributions* 10(5–6):447–459.

Upton, G. J. G., and B. Fingleton. 1985. *Spatial Data Analysis by Example. Vol. 1, Point Pattern and Quantitative Data.* New York: John Wiley & Sons.

Volk, W. 1980. *Applied Statistics for Engineers.* Huntington, NY: Krieger Publishing Company.

Williams, C. K. I. 1998. Prediction with Gaussian processes: From linear regression to linear prediction and beyond. In *Learning in Graphical Models,* M. I. Jordan, ed., 599–612. Cambridge, MA: MIT Press.

Williams, M. S. 1997. A regression technique accounting for heteroscedastic and asymmetric error. *Journal of Agriculture, Biology and Environmental Statistics* 2:108–129.

Youden, W J. 1951. *Statistical Methods for Chemists.* New York: John Wiley & Sons.

# 5

## R Statistical Package

This chapter is based on the link provided by the R Project (http://www.r-project.org/); if you are writing R Extensions visit http://CRAN.R-project.org/doc/Rnew; for manuals, go to http://CRAN.R-project.org/manuals.html; looking for frequently asked questions, you may look to http://CRAN.R-project.org/faqs.html; and if you need help use http://www.R-project.org/posting-guide.html.

There are other contributions by education institutes around the world if you have concern about the R Statistical Package in general. Also, material provided within this text were cited and modified from the workshop by Dr. Hank Stevens, "R for Statistics, By Example: A Workshop for the Ecological Society of America (ESA)," August 5–10, 2006, Memphis, TN, USA.

When you start with R from the R Project Web site (http://www.r-project.org/), you will see the following:

> R version 2.4.0 (2006-10-03) (This is the latest version of "R")
> Copyright (C) 2006 the R Foundation for Statistical Computing
> ISBN 3-900051-07-0
> R is free software and comes with ABSOLUTELY NO WARRANTY.
> Type 'license()' or 'licence()' for distribution details.
> Natural language support but running in an English locale.
> R is a collaborative project with many contributors around the globe.
> Type 'contributors()' for more information and 'citation()' on how to cite
> R or R packages in publications.
> Type 'demo()' for some demos, 'help()' for on-line help, or 'help.start()'
> for an HTML browser interface to help. Type 'q()' to quit R.
>
> > license()

R software is distributed under the terms of the GNU GENERAL PUBLIC LICENSE Version 2, June 1991. The terms of this license are in a file called COPYING which you should have received with this software.

If you have not received a copy of this file, you can obtain one via WWW at http://www.gnu.org/copyleft/gpl.html, or by writing to: The Free Software Foundation, Inc., 51 Franklin Street, Fifth Floor, Boston, MA 02110-1301, USA.

A small number of files (the API header files and export files, listed in R_HOME/COPYRIGHTS) are distributed under the LESSER GNU GENERAL PUBLIC LICENSE version 2.1. This can be obtained via WWW at http://www.gnu.org/copyleft/lgpl.html, or by writing to the address above.

## Overview of R Statistics (R)

S is a quantitative programming environment developed at AT&T Bell Labs in the 1970s. S-Plus was begun in the 1980s, and R was begun in the 1990s by Robert Gentleman and Ross Ihaka of the Statistics Department of the University of Auckland. Nearly over 30 senior statisticians provide the core development group of the R language, including the primary developer of the original S language, John Chambers, of Bell Labs.

### What Is R?

1. R is a language and environment for statistical computing and graphics.
2. R is a GNU project that is similar to the S language, a different implementation of S.
3. R provides a very wide variety of statistical (linear and nonlinear modeling, classical statistical tests, time-series analysis, classification, clustering, etc.) and graphical techniques, and is highly extensible.
4. S language is often the vehicle of choice for research in statistical methodology, and R provides an Open Source route to participation in that activity.

### Strengths of R/S

- We can learn and teach R both simply and quickly, and can accomplish a lot with very little code.
- Friendly. You can extend the functionality of R by writing code. This may be a simple function for personal use or a whole new family of statistical procedures in a new package.
- Much variety of statistical and computing procedures. This derives from the ease with which R/S can be extended and shared by users around the world.
- Rapid updates with R.
- All data analyses should be well documented and validated, and this only happens reliably when the analyses are performed with scripts or programs, as in R or SAS or S-Plus.
- Getting help from other people is easy. Any scripted language can be quickly and easily shared with someone who can help you or go to the R Web site.

- Writing code allows you to do anything you want a huge number of times. It also allows very simple updates with new data.

- Plots can be produced, including mathematical symbols and formulae where needed. Great care has been taken over the defaults for the minor design choices in graphics, but the user retains full control.

- R is available as free software under the terms of the Free Software Foundation's GNU General Public License in source code form. It compiles and runs out of the box on a wide variety of UNIX platforms and similar systems (including FreeBSD and Linux). It also compiles and runs on Windows 9x/NT/2000 and Mac OS.

- Accessible. Try this: Type *R* into Google. The R Project page is the first noncommercial hit.

### The R Environment

R is an integrated suite of software facilities for data manipulation, calculation, and graphical display. It includes

- An effective data handling and storage facility

- A suite of operators for calculations on arrays, in particular matrices

- A large, coherent, integrated collection of intermediate tools for data analysis, graphical facilities for data analysis and display either on-screen or on hardcopy, and a well-developed, simple and effective programming language, which includes conditionals, loops, user-defined recursive functions, and input and output facilities

The term *environment* is intended to characterize it as a fully planned and coherent system, rather than an incremental accretion of very specific and inflexible tools, as is frequently the case with other data analysis software. R, like S, is designed around a true computer language, and it allows users to add additional functionality by defining new functions. Much of the system is itself written in the R dialect of S, which makes it easy for users to follow the algorithmic choices made. For computationally intensive tasks, C, C++, and Fortran code can be linked and called at run time. Advanced users can write C code to manipulate R objects directly.

Many users think of R as a statistics system; however, think of it as an environment within which statistical techniques are implemented. R can be extended (easily) via packages. There are about eight packages supplied with the R distribution and many more are available through the CRAN family of Internet sites covering a very wide range of modern statistics. R has its own LaTeX-like documentation format, which is used to supply comprehensive documentation, both on-line in a number of formats and in hardcopy.

### Scripts

- Scripts are text files that contain your analysis, that is, they contain both code to do stuff, and comments about what you are doing and why.
- Scripts are not Microsoft Word documents that require Microsoft Word to open them, but rather, simple text files one could open in Notepad or SimpleText.
- Scripts contain code with which to perform your data analysis.
- Scripts are a written record of everything you do along the way to achieving your results.
- Scripts are the core of data analysis, and provide many of the benefits of using a command-driven system, whether R, S-Plus, or SAS.

### Working with R on Your COMPUTER

Create a new directory (i.e., a folder called "myR") in a convenient place. Load all of the provided files into that directory and open the R GUI in a manner that is appropriate for your operating system and setup. R will begin by opening its primitive GUI, and provide a prompt. In the R GUI, the prompt is >. You will type comments and code into the R scripts provided, thus adding your comments and code to the script.

### Begin to Use R

Comments that R will ignore are preceded by #.

```
# I am doing regression analyzing on my data
```

Submit your first command as:

```
help.start() # Remember - Windows: type Ctrl-R. Macs: select
and # hit return.
```

Find out where you are ("GET the Working Directory")

```
getwd()
```

Set the working directory by typing the complete path name to "myR" in this script, here, now, and submitting the code with Ctrl-R or Command return. In R, the path name always uses forward slashes, like UNIX (/), not back slashes like DOS. Your path name for "myR" might look like one of these.

```
> getwd()
```

```
[1] "C:/myR"
```

```
>setwd("C:/myR")
```

Remember:

- All text to submit to R is in typewriter (i.e., Courier or Times New Roman) font.
- Text that begins with # will be ignored by R.
- Text that does not make sense to R will cause R to return an error message, but will not otherwise mess things up.
- To add a comment, simply type one or more #, followed by your comment.

## Statistical Analysis Examples Using R

Open R and change directory (your desire): In RGUI, click on file, *change dir*, window will open, choose working directory *browse to find your desire directory* and hit enter or return key.

```
> getwd()
```

```
[1] "C:/myR"
```

```
>setwd("C:/myR"
```

Now we can work on your data or other data.

### Common Statistics

- Mean, median (mean, median)
- Variance, standard deviation, range (var, sd, range)
- Minimum, maximum (min, max)
- Quantiles (summary, quantile)
- t-based confidence interval (relies on qt, length, sd)

### Common Graphics

- Stem and leaf display (stem)
- Histogram (hist)
- Enclosing a graph in box (box)
- Adding straight lines by formula (abline)
- Adding error bars (arrows)
- Adding points (points)
- Making graphics files (e.g., pdf, png, postscript, jpeg, bitmap)

## Common Programming

- Help (?, help)
- Assignment (<-, =)
- Simple data entry (c)
- Arithmetic (+, − , _, /, %%,%/%)
- Logarithms (log, log2, log10, log(x, base=y)
- Describing objects (length, dim)
- Extraction ([ ])
- Adding error bars (arrows)

Note: Take copies about what you did above! Have Fun!
Check your working directory!

```
> getwd()

[1] "C:/myR"

> setwd("C:/myR")

Enter data using c()

> myheight<- c(5.1,10.5, 6.3,4.1,5.8,6.7,7.1,3.2,5.9,5.0,6.0,8
.1,NA)
> length(myheight)

[1] 13

List data

> myheight

 [1] 5.1 10.5 6.3 4.1 5.8 6.7 7.1 3.2 5.9 5.0 6.0 8.1 NA

> mean(myheight, na.rm= TRUE)

[1] 6.15

> is.na(myheight) # Tests whether each element of myheight is
missing (is NA)

 [1] FALSE FALSE FALSE FALSE FALSE FALSE FALSE FALSE FALSE
FALSE FALSE FALSE
[13] TRUE

> length( myheight[is.na(myheight)])

[1] 1
```

### Create and Examine a Logical Vector

```
> !is.na(myheight)
```

```
[1] TRUE TRUE TRUE TRUE TRUE TRUE TRUE TRUE TRUE TRUE TRUE TRUE
[13] FALSE
```

```
> myheight<-myheight[ !is.na(myheight)] #This will let ask
which is true when using [ ]
```

```
dim(myheight) # Returns NULL
```

### Working on Graphical Display of Data (Data Distributions)

```
> stem(myheight) # How to interpret the stem and leaf plot?
```

The decimal point is at the |.

```
 2 | 2
 4 | 10189
 6 | 0371
 8 | 1
10 | 5
```

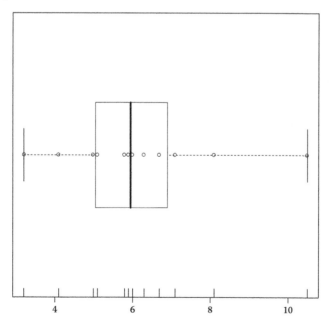

**FIGURE 5.1**
Graphical display using boxplot for "myheight" data.

```
> boxplot(myheight)
> boxplot(myheight, range =0, horizontal=TRUE)
> rug(myheight) #small tick mark were data are!
> points(myheight, rep(1,length(myheight)))
```

### Develop a Histogram

```
> points(myheight, rep(1,length(myheight)))
> hist(myheight)
> rug(myheight)
```

### Data Comparison between the Data and an Expected Normal Distribution

```
> qqnorm(myheight)
> qqline(myheight)
> shapiro.test(myheight)
```

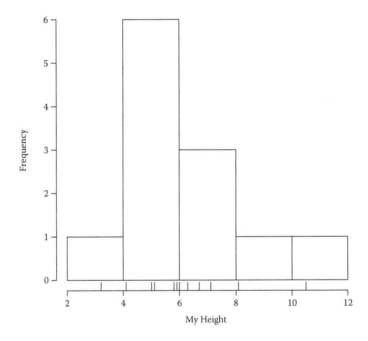

**FIGURE 5.2**
Using histogram for displaying "myheight" data.

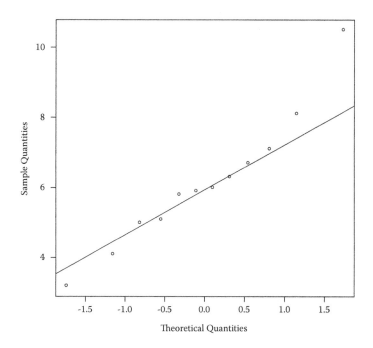

**FIGURE 5.3**
Normal distribution plot using Q-Q-plot for "myheight" data.

```
 Shapiro-Wilk normality test

data: myheight
W = 0.9492, p-value = 0.6259
```

Find out if the transformation on data will help or not.

```
> qqnorm(log(myheight))
> qqline(log(myheight))
> shapiro.test(log(myheight))

    Shapiro-Wilk normality test

data: log(myheight)
W = 0.979, p-value = 0.9791
```

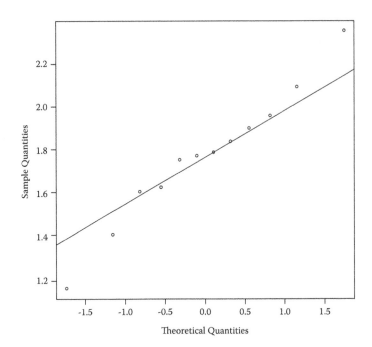

**FIGURE 5.4**
Transformation of the data using "log(myheight)" data.

What is your answer?

## More Statistical Analysis

```
> sd.myheight<-sd(myheight)

[1] 1.894730

> n<-length(myheight)
> n

[1] 12

> se.myheight<-sd.myheight/sqrt(n)
> se.myheight

[1] 0.5469613
```

## Reading New Variable (Enter new data set, WEIGHT)

```
> myweight<-c(100,90,50,200,150,160,250,180,170,190,174,130)
```

```
> mean(myweight)

[1] 153.6667

> n<-length(myweight)
> n

[1] 12

> sd.myweight<-sd(myweight)
> sd.myweight

[1] 54.28432

> se.myweight<-sd.myweight/sqrt(n)
> se.myweight

[1] 15.67053

> cor(myheight,myweight)

[1] -0.3602631
```

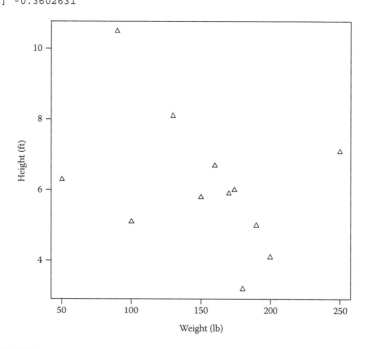

**FIGURE 5.5**
Plotting the relationship between weight and height from "myweight" and "myheight" data, respectively.

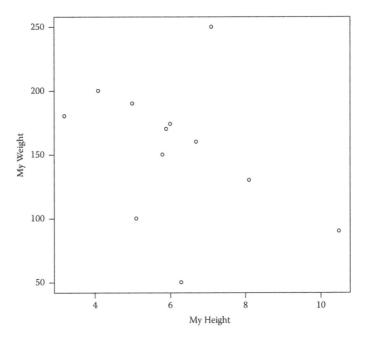

**FIGURE 5.6**
Simple graph to plot height vs. weight.

## Plotting Weight and Height

```
> plot(myweight, myheight)
> plot(myweight, myheight, pch=2, col=1)
> plot(myweight, myheight,main= "Relationship Between Weight
and Height", ylab="Height(Ft)", xlab="Weight lb", pch=2, col=1)
```

## Test of Association

Test for linear correlation between height and weight.

```
> cor(myheight, myweight)

[1] -0.3602631

> mycor<-cor.test(myheight,myweight)
> mycor

    Pearson's product-moment correlation

data: myheight and myweight
t = -1.2213, df = 10, p-value = 0.25
```

alternative hypothesis: true correlation is not equal to 0
95 percent confidence interval:
 -0.7741126 0.2693224

sample estimates:
     cor
-0.3602631

> ?cor.test (See help on cor!)

> mycor1<-cor.test(myheight,myweight, method="spearman")

#Note if you type "Spearman" instead of "spearman", we get get
error this due to the fcat, we used upper case "S"

> mycor1

     Spearman's rank correlation rho

data: myheight and myweight
S = 408, p-value = 0.1689

alternative hypothesis: true rho is not equal to 0

sample estimates:
    rho
-0.4265734

## Some Basic Regression Analysis

```
> plot(myweight ~ myheight)
> myhtwt.r<-lm(myheight ~ myweight) {Develop Regression Model}
> myhtwt.r
```

Call:
lm(formula = myheight ~ myweight)

Coefficients:
(Intercept)    myweight
   8.08229     -0.01257

```
> abline(myhtwt.r)
> layout(matrix(1:2, nr=2))
> par(cex=1.2)
> plot(myhtwt.r, which = 1:2)
> summary(myhtwt.r)
```

Call:
lm(formula = myheight ~ myweight)

Residuals:
     Min       1Q    Median       3Q       Max
  -2.6189   -1.2320   -0.2204   0.8853    3.5494

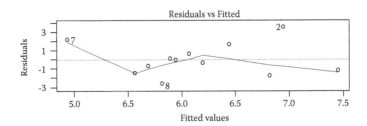

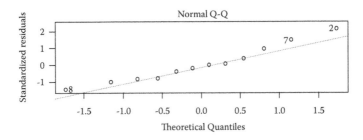

**FIGURE 5.7**
Plotting the residuals and standardized residuals for estimate regression mode between weight and height.

```
Coefficients:
      Estimate Std. Error t value Pr(>|t|)
(Intercept)    8.08229    1.67026    4.839    0.000683    ***
myweight      -0.01257    0.01030   -1.221    0.249999
---
Signif. codes: 0 '***' 0.001 '**' 0.01 '*' 0.05 '.' 0.1 ' ' 1

Residual standard error: 1.854 on 10 degrees of freedom
Multiple R-Squared: 0.1298,    Adjusted R-squared: 0.04277
F-statistic: 1.491 on 1 and 10 DF, p-value: 0.25
```

## Case Study

Now we are ready to work on real data. The data was collected by Dr. Mohammed A. Kalkhan's research team at Natural Resource Ecology Laboratory, Colorado State University during summer field seasons 2003 and 2004 using pixel nested plot (PNP) design within the eastern region of Rocky Mountain National Park, Colorado (see Kalkhan et al. 2007a, 2007b).

For this exercise, we will use data for 122 PNPs of 15 m × 15 m. Your task is to develop a plant diversity geospatial model and map using R Software by following these steps:

1. Read the data (use the text file in your home directory "myR" with name "romo-tplant.txt").
2. Test for linear correlation (between x, y, total plant).
3. Develop an inverse weight matrix (using x and y coordinates).
4. Test for the spatial autocorrelation using Moran's *I* and Geary's *C*.
5. Test for cross-correlation statistics.
6. Comment about your findings.

Here are the steps to work on this exercise: Click on the R icon on your computer; the following will appear:

R version 2.4.0 (2006-10-03)

Copyright (C) 2006 The R Foundation for Statistical Computing

ISBN 3-900051-07-0

R is free software and comes with ABSOLUTELY NO WARRANTY.

You are welcome to redistribute it under certain conditions.

Type 'license()' or 'licence()' for distribution details.

Natural language support but running in an English locale

R is a collaborative project with many contributors.

Type 'contributors()' for more information and

'citation()' on how to cite R or R packages in publications.

Type 'demo()' for some demos, 'help()' for on-line help, or

'help.start()' for an HTML browser interface to help.

Type 'q()' to quit R.

[Previously saved workspace restored]

```
> choose CRAN mirror()
```

To load the spatial library, you need to type "RGui" window environment:

```
> Spatial_start() {This will start the spatial library to
perform the following commend lines}
[1] "C:/myR/R/RSpatial/"
> romoplant<-spinput("c:/myr/romo-tplant.txt",T)
```

```
> romoplant[1,]

        xutm              yutm          tplant         band1          band2
  446811.00000    4472993.00000      35.50000      73.00000      65.00000
        band3             band4          band5         band6h         band6l
     68.00000         76.00000     114.00000     175.00000     145.00000
        band7             band8            elv         absasp           slp
     75.00000         66.00000    2623.00000      23.19859      12.52359
     landform
     -2.40000

> cor(romoplant[,1:3])  {Test for linear correlation}
               xutm              yutm          tplant
xutm        1.0000000     -0.2308289     -0.1502507
yutm       -0.2308289      1.0000000      0.0815509
tplant     -0.1502507      0.0815509      1.0000000

> romoplant.wt<-spwtdist(romoplant[,1],romoplant[,2],
a=1,band=0, binary=F, rescale=F)

[romoplant[,26, romoplant[,27]: reading x and y values from
your data;
a= 1 mean use inverse distance; band= 0: mean no restriction
on weight being set =0; binary = F, do not use binary weight;
and resacle = F= do not resacle]

----------------------------------------------------------------

Characteristics of distance matrix
Dimension: 122
Average distance between points: 9019.244
Distance range:        27000.23
Minimum distance between points: 152.5025
Quartiles

   First:        5237.22
   Median:       8242.161
   Third:       12128.63

Maximum distance between points: 27152.73
Create weight matrix based on "Inverse Distance Function" and
rescale using inverse distance = 1)

> romoplant.wt<-spwtdist(romoplant[,1],romoplant[,2],
a=1,band=0, binary=F, rescale=T)

Characteristics of distance matrix
Dimension: 122
Average distance between points: 59.14163
Distance range: 177.0478
Minimum distance between points: 1
Quartiles
```

```
First:       34.34187
Median:      54.04609
Third:       79.5307
```

Maximum distance between points: 178.0478

## Test for Spatial Autocorrelation Using Moran's *I*

```
> morani(romoplant[,1], w=romoplant.wt)
```

```
UNDER NORMAL APPROXIMATION
Moran's I is     = 0.317939
Mean of I is     = -0.008264
St. Dev of I     = 0.013282
Z-Value          = 24.55935
P-Value(2-side)  = 0
```

```
UNDER RANDOMIZATION ASSUMPTION
```

```
Moran's I is     = 0.317939
Mean of I is     = -0.008264
St. Dev of I     = 0.013257
Z-Value          = 24.60598
P-Value(2-side)  = 0
```

```
> morani(romoplant[,2], w=romoplant.wt)
```

```
UNDER NORMAL APPROXIMATION
Moran's I is     = 0.328899
Mean of I is     = -0.008264
St. Dev of I     = 0.013282
Z-Value          = 25.38451
P-Value(2-side)  = 0
```

```
UNDER RANDOMIZATION ASSUMPTION
Moran's I is     = 0.328899
Mean of I is     = -0.008264
St. Dev of I     = 0.013318
Z-Value          = 25.31572
P-Value(2-side)  = 0
```

```
> morani(romoplant[,3], w=romoplant.wt)
```

```
UNDER NORMAL APPROXIMATION
Moran's I is     = -0.021661
Mean of I is     = -0.008264
St. Dev of I     = 0.013282
Z-Value          = -1.008625
P-Value(2-side)  = 0.313155
```

```
UNDER RANDOMIZATION ASSUMPTION
Moran's I is     = -0.021661
Mean of I is     = -0.008264
St. Dev of I     = 0.01326
Z-Value          = -1.010306
P-Value(2-side)  = 0.312349
```

## Test for Spatial Autocorrelation Using Geary's C

```
> gearyc(romoplant[,1], w=romoplant.wt)

UNDER NORMAL APPROXIMATION

Geary's C is     = 0.566727
Mean of C is     = 1
St. Dev of C     = 0.019345
Z-Value          = -22.39684
P-Value(2-side)  = 0

UNDER RANDOMIZATION ASSUMPTION

Geary's C is     = 0.566727
Mean of C is     = 1
St. Dev of C     = 0.020495
Z-Value          = -21.14036
P-Value(2-side)  = 0

> gearyc(romoplant[,2], w=romoplant.wt)

UNDER NORMAL APPROXIMATION
Geary's C is     = 0.571031
Mean of C is     = 1
St. Dev of C     = 0.019345
Z-Value          = -22.17436
P-Value(2-side)  = 0

UNDER RANDOMIZATION ASSUMPTION
Geary's C is     = 0.571031
Mean of C is     = 1
St. Dev of C     = 0.017561
Z-Value          = -24.42719
P-Value(2-side)  = 0

> gearyc(romoplant[,3], w=romoplant.wt)

UNDER NORMAL APPROXIMATION
Geary's C is     = 1.023527
Mean of C is     = 1
```

```
St. Dev of C     = 0.019345
Z-Value          = 1.216159
P-Value(2-side)  = 0.223924


UNDER RANDOMIZATION ASSUMPTION
Geary's C is     = 1.023527
Mean of C is     = 1
St. Dev of C     = 0.020358
Z-Value          = 1.155633
P-Value(2-side)  = 0.247831
```

### Test for Spatial Cross-Correlation Using Bi-Moran's *I*

```
> bimorani(romoplant[,13],romoplant[,2], w=romoplant.wt)

UNDER RANDOMIZATION ASSUMPTION
Moran's I is     = 0.026045
Mean of I is     = 4.9e-05
St. Dev of I     = 0.009424
Z-Value          = 2.75849
P-Value(2-side)  = 0.005807


> bimorani(romoplant[,1],romoplant[,2], w=romoplant.wt)

UNDER RANDOMIZATION ASSUMPTION
Moran's I is     = -0.095433
Mean of I is     = 0.001908
St. Dev of I     = 0.009655
Z-Value          = -10.08202
P-Value(2-side)  = 0


> bimorani(romoplant[,1],romoplant[,3], w=romoplant.wt)

UNDER RANDOMIZATION ASSUMPTION
Moran's I is     = -0.052473
Mean of I is     = 0.001242
St. Dev of I     = 0.009566
Z-Value          = -5.614897
P-Value(2-side)  = 0


> bimorani(romoplant[,2],romoplant[,3], w=romoplant.wt)

UNDER RANDOMIZATION ASSUMPTION
Moran's I is     = 0.016781
Mean of I is     = -0.000674
St. Dev of I     = 0.009488
Z-Value          = 1.839605
P-Value(2-side)  = 0.065826
```

```
> ls()
[1] "a"              "bimorani"       "ols"           "romoplant"
[5] "romoplant.wt"   "S_HOME"     "Spatial_start"    "stepwise"
[9] "test"
>
```

End of this exercise: From what you learn, create your own report to remember what you worked on.

---

## Trend Surface Analysis

Now we need to develop a regression model to measure coarse scale variability. In other words develop trend surface analysis using OLS, or spatial AR.

```
Run OLS Procedure:

Using all independent variables to predict total plants:

> plant.ols<-ols(plant[,1],plant[,c(2:14)],w=romoplant.wt)

Residual Standard Error=23.4021
R-Square=0.3697
F-statistic (df=13, 108)=4.8723
p-value=0
```

|           | Estimate | Std.Err  | t-value | Pr(>\|t\|) |
|-----------|----------|----------|---------|-----------|
| Intercept | 204.2839 | 202.8964 | 1.0068  | 0.3163    |
| band1     | 0.0718   | 1.1644   | 0.0617  | 0.9509    |
| band2     | -0.8806  | 1.6391   | -0.5372 | 0.5922    |
| band3     | -0.2269  | 0.9195   | -0.2468 | 0.8056    |
| band4     | 1.2782   | 0.4366   | 2.9275  | 0.0042    |
| band5     | 0.0163   | 0.2003   | 0.0813  | 0.9353    |
| band6h    | 1.6757   | 2.2751   | 0.7366  | 0.4630    |
| band6l    | -3.3110  | 4.0888   | -0.8098 | 0.4198    |
| band7     | 0.0638   | 0.0634   | 1.0070  | 0.3162    |
| band8     | 0.5215   | 0.6391   | 0.8160  | 0.4163    |
| elv       | -0.0063  | 0.0106   | -0.5970 | 0.5518    |
| absasp    | -0.0155  | 0.0417   | -0.3715 | 0.7110    |
| slp       | -0.1864  | 0.2199   | -0.8478 | 0.3985    |
| landform  | 4.5004   | 1.4200   | 3.1693  | 0.0020    |

```
Log(like)      = -550.3198
AIC            = 1128.640
AICC           = 1132.565
Schwartz       = 1167.896
```

```
Moran's I (res)     = -0.0169
Mean of I           = -1720.485
Std Dev of I        = 2813.435
Z-value of I        = 0.6115
P-value(2-side)     = 0.5409


Lagrange Mult       = 0.9029
P-value(2-side)     = 0.342
```

Using only the significant independent variables from first OLS Run:

```
> plant.ols<-ols(plant[,1],plant[,c(5,14)],w=romoplant.wt)
```

```
Residual Standard Error=24.5569
R-Square=0.2352
F-statistic (df=2, 119)=18.3015
p-value=0
```

|  | Estimate | Std.Err | t-value | Pr(>\|t\|) |
|---|---|---|---|---|
| Intercept | -9.3922 | 11.4997 | -0.8167 | 0.4157 |
| band4 | 0.9284 | 0.1777 | 5.2256 | 0.0000 |
| landform | 4.9958 | 1.3296 | 3.7573 | 0.0003 |

```
Log(like)           = -562.1131
AIC                 = 1130.226
AICC                = 1130.430
Schwartz            = 1138.638

Moran's I (res)     = -0.0144
Mean of I           = -0.0171
Std Dev of I        = 0.0196
Z-value of I        = 0.1393
P-value(2-side)     = 0.8892

Lagrange Mult       = 0.659
P-value(2-side)     = 0.4169
```

The Moran's *I* residuals and Lagrange show that no spatial autocorrelation pattern exists or that it is not significant. We can develop geospatial model map.

```
> plant.ols$coef
Intercept           band4              landform
-9.3921670          0.9283845          4.9958322
```

In ARC using Grid or ARCMAP using raster calculator, you can write the final model as:

```
Tplant: -9.3921670 + 0.9283845*band4 + 4.9958322*landform (The
output is grid file)
> summary(plant.ols$resid)
 Min. 1st Qu. Median Mean 3rd Qu. Max.
-4.953e+01 -2.010e+01 -4.304e+00 5.330e-16 1.629e+01 7.305e+01
```

Plot residuals and predicted values for total plants:

```
> par(mfrow=c(2,1))
> hist(plant.ols$resid,xlab="Residulas",ylab="Frequency")
> plot((plant[,1]- plant.ols$resid),plant.
ols$resid,xlab="Predicted Total Plants",ylab="Residulas")
```

## Test for Spatial Autocorrelation of the Residuals

```
> cor(plant.ols$resid,romoplant.wt%*%plant.ols$resid)
 [,1]
[1,] -0.1812981
```

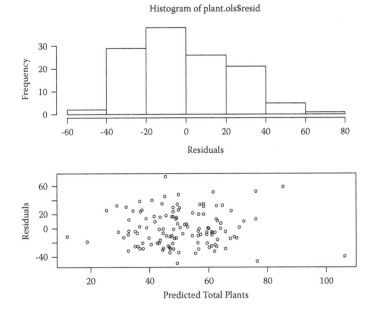

**FIGURE 5.8**
Histogram and plot of residuals and predicted values of total plants at Rocky Mountain National Park, Colorado.

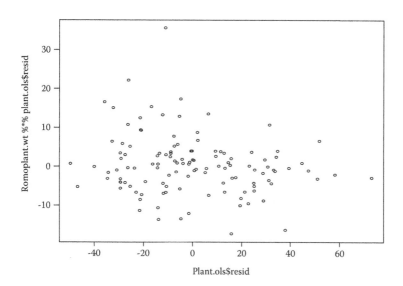

**FIGURE 5.9**
Plot of spatial autocorrelation test for the residuals.

```
> plot(plant.ols$resid,romoplant.wt%*%plant.ols$resid)
```

## Test for Moran's *I* for Residuals

```
> morani(plant.ols$resid, w=romoplant.wt)

UNDER NORMAL APPROXIMATION
Moran's I is      = -0.014412
Mean of I is      = -0.008264
St. Dev of I      = 0.013282
Z-Value           = -0.462871
P-Value(2-side)   = 0.643457

UNDER RANDOMIZATION ASSUMPTION
Moran's I is      = -0.014412
Mean of I is      = -0.008264
St. Dev of I      = 0.013296
Z-Value           = -0.46241
P-Value(2-side)   = 0.643787
```

From the all steps you worked with, what is your conclusion and what do you think?

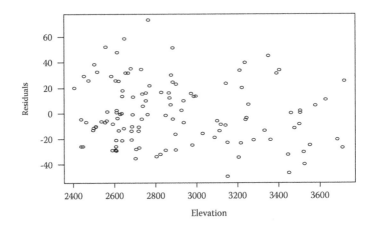

**FIGURE 5.10**
Plot the residuals for total plants over the elevation.

## Using Spatial AR Model without Regression

```
> spatar1(plant.ols$resid,w=romoplant.wt,xreg=F)


Residual Standard Error = 24.0628 , Multiple R-Square = 0.0156
N = 122, F-Statistic = 0.9687 on 2 and 120 df, p-value = 0.3825
                 coef          std.err        t.stat        p.value
Intercept        0.0125        1.3200         0.0095        0.9925
lambda          -0.6504        0.5713        -1.1385        0.2572
Variance       579.0198       74.4174         7.7807        NA

Log(like)            = -561.5663
AIC                  = 1127.132
AICC                 = 1127.233
Schwartz             = 1132.741
                                   Value        P-Value
Likelihood Ratio Test (df=1)       1.0937       0.2957
```

## Using Spatial AR with Regression (Using All Independent Variables as with OLS Model)

```
>plant.sar<-spatar1(plant[,1],plant[,c(2:14)],w=romoplant.wt)


Residual Standard Error = 21.7254 , Multiple R-Square = 0.3863
N = 122 , F-Statistic = 5.1204 on 15 and 107 df, p-value = 0
```

|          | coef     | std.err  | t.stat  | p.value |
|----------|----------|----------|---------|---------|
| Intercept | 177.6107 | 189.5554 | 0.9370  | 0.3509  |
| band1    | 0.2151   | 1.0741   | 0.2002  | 0.8417  |
| band2    | -0.8526  | 1.5205   | -0.5607 | 0.5762  |
| band3    | -0.3124  | 0.8399   | -0.3720 | 0.7106  |
| band4    | 1.3011   | 0.4060   | 3.2045  | 0.0018  |
| band5    | 0.0107   | 0.1876   | 0.0568  | 0.9548  |
| band6h   | 1.6068   | 2.1577   | 0.7447  | 0.4581  |
| band6l   | -3.1269  | 3.8746   | -0.8070 | 0.4214  |
| band7    | 0.0670   | 0.0592   | 1.1316  | 0.2604  |
| band8    | 0.4238   | 0.5963   | 0.7107  | 0.4788  |
| elv      | -0.0032  | 0.0090   | -0.3614 | 0.7185  |
| absasp   | -0.0122  | 0.0386   | -0.3160 | 0.7526  |
| slp      | -0.1502  | 0.1953   | -0.7691 | 0.4435  |
| landform | 4.6006   | 1.3273   | 3.4661  | 0.0008  |
| lambda   | -0.9002  | 0.5954   | -1.5119 | 0.1335  |
| Variance | 471.9945 | 60.8588  | 7.7556  | NA      |

```
Log(like)    = -549.4441
AIC          = 1128.888
AICC         = 1133.417
Schwartz     = 1170.948
```

|                                | Value  | P-Value |
|--------------------------------|--------|---------|
| Likelihood Ratio Test (df=1)   | 1.7513 | 0.1857  |

```
> plant.sar<-spatar1(plant[,1],plant[,c(5,14)],w=romoplant.wt)
```

```
Residual Standard Error = 24.0492 , Multiple R-Square = 0.248
N = 122 , F-Statistic = 10.0605 on 4 and 118 df, p-value = 0
```

|          | coef     | std.err | t.stat  | p.value |
|----------|----------|---------|---------|---------|
| Intercept | -6.8077  | 10.9165 | -0.6236 | 0.5341  |
| band4    | 0.8876   | 0.1711  | 5.1882  | 0.0000  |
| landform | 4.9858   | 1.3152  | 3.7909  | 0.0002  |
| lambda   | -0.6844  | 0.5750  | -1.1902 | 0.2364  |
| Variance | 578.3663 | 74.3623 | 7.7777  | NA      |

```
Log(like)    = -561.5388
AIC          = 1131.078
AICC         = 1131.419
Schwartz     = 1142.294
```

|                                | Value  | P-Value |
|--------------------------------|--------|---------|
| Likelihood Ratio Test (df=1)   | 1.1486 | 0.2838  |

```
> summary(plant.sar$resid)
 Min. 1st Qu. Median Mean 3rd Qu. Max.
-4.913e+01 -1.886e+01 -3.335e+00 4.942e-16 1.614e+01 7.229e+01
```

```
> plant.sar$coef
```

| Intercept   | band4      | landform   |
| ----------- | ---------- | ---------- |
| -6.8077015  | 0.8875747  | 4.9857981  |

### Analysis of Residuals

```
> plant.pred<-plant[,1]-plant.sar$resid
> par(mfrow=c(2,2))
> hist(plant.sar$resid,ylab="Frequency",xlab="Residuals")
> plot((plant[,1]-plant.sar$resid),plant.sar$resid,ylab="Resid
uals",xlab="Predicted Total Plants")

> qqnorm(plant.sar$resid)
> qqline(plant.sar$resid)
> plot(plant.pred,plant[,1],ylab="Observed Total
Plants",xlab="Predicted Total Plants")
> par(mfrow=c(1,1))
> plot(plant[,1],plant.sar$resid)
```

From the all steps you worked with, what is your conclusion and what do you think?

### Develop Variogram Model (Modeling Fine Scale Variability)

```
> plant.1st<-list(romoplant[,1],romoplant[,2],plant.ols$resid,
xl=422582,15,xu=458159,yl=4445263,yu=4489250
```

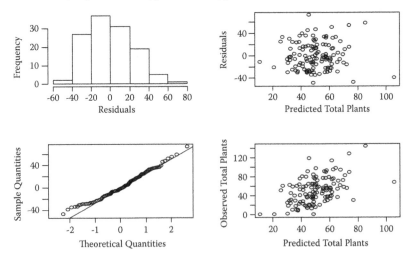

**FIGURE 5.11**
Histogram and plot of residuals for predicting total plants.

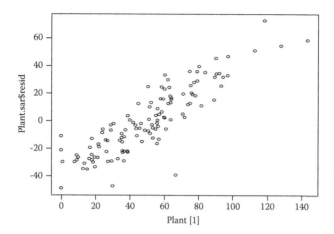

**FIGURE 5.12**
Plot dependent variable vs. residuals spatial AR model.

```
> plant.var<-variogrm(romoplant[,1],romoplant[,2],plant.
ols$resid,15,dmax=0)

> plant.var
$x
 [1]  0.000 2399.553 4799.106 7198.659 9598.212 11997.765
14397.317 16796.871 19196.424 21595.977 23995.529

$y
 [1]  0.0000 473.0859 627.7654 614.7528 564.8698 586.4839
601.3871 556.2465 552.1041 603.1296 636.4980

$ni
 [1]  649 1058 1444 1423 1051 725 515 333 177 88 35

$type
 [1] «var»

> plant.exp<-fitvar(plant.var,0,600,25000,model=»exp»,wt=T)

Least Squares Estimate

Nugget          = 0
Sill            = 636.1319
Range           = 4051.067
alpha           = 0
s.e.            = 25.22166

Log(like)       = -110.1178
AIC             = 226.2355
```

```
AICC                    = 229.6641
Schwartz                = 227.4292

> plant.gau<-fitvar(plant.var,0,600,25000,model="gau",wt=T)

Least Squares Estimate

Nugget                  = 0
Sill                    = 599.0063
Range                   = 2853.601
alpha                   = 0
s.e.                    = 24.47461

Log(like)               = -132.1708
AIC                     = 270.3416
AICC                    = 272.5234
Schwartz                = 272.4658

> plant.sph<-fitvar(plant.var,0,600,25000,model=»sph»,wt=T)

Least Squares Estimate

Nugget                  = 0
Sill                    = 579.4583
Range                   = 672.7351
alpha                   = 0
s.e.                    = 24.07194
```

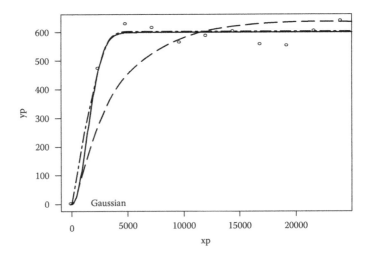

**FIGURE 5.13**
Variogram models for fine scale variability to predict total plants.

```
Log(like)           = -101.9561
AIC                 = 209.9121
AICC                = 213.3407
Schwartz            = 211.1058
```

**Plotting Variogram Model**

```
> plant.var<-variogrm(romoplant[,1],romoplant[,2],plant.
ols$resid,15,dmax=0)
> lines(sphervar(x,plant.sph),lty=6,lwd=3)
> lines(expvar(x,plant.exp),lty=5,lwd=3)
> lines(gauvar(x,plant.gau),lty=1,lwd=3)
> legend(.3,.2,legend=c("Gaussian","Exponential","Spherical",l
ty=c(6,5,1)))
> title("Variogarm Model")
```

# References

Crawley, M. J. 2002. *Statistical Computing: An Introduction to Data Analysis Using S-Plus.* New York: John Wiley & Sons.

Dalgaard, P. 2002. *Introductory Statistics with R.* New York: Springer-Verlag.

Everitt, B. S., and T. Hothorn. 2006. *A Handbook of Statistical Analysis Using R.* New York: Chapman & Hall/CRC.

Faraway, J. J. 2006. *Extending the Liner Model with R Generalized Liner Mixed Effects and Nonparamterics Regression Models.* Boca Raton, FL: Taylor & Francis.

Kalkhan, M. A., E. J. Stafford, and T. J. Stohlgren. 2007a. Rapid plant diversity assessment using a pixel nested plot design: A case study in Beaver Meadows, Rocky Mountain National Park, Colorado, USA. *Diversity and Distributions* 13:379–388.

Kalkhan, M. A., E. J. Stafford, P. J. Woodly, and T. J. Stohlgren. 2007b. Exotic plant species invasion and associated abiotic variables in Rocky Mountain National Park, Colorado, USA. *Journal of Applied Soil Ecology* 37:622–634.

Pinheiro, J., and D. Bates. 2000. *Mixed-effects Models in S and S-PLUS.* New York: Springer.

The R Project for Statistical Computing, http://www.r-project.org/.

Venables, W. W., and B. D. Ripley. 2000. *S Programming.* New York: Springer-Verlag.

Venables, W.W., and B. D. Ripley 2002. *Modern Applied Statistics with S-Plus.* New York: Springer-Verlag.

Verzani, J. 2005. *Using R for Introductory Statistics.* Boca Raton, FL: Chapman & Hall/CRC.

Wood, S. N. 2006. *Generalized Additive Models: An Introduction with R.* Boca Raton, FL: Chapman & Hall/CRC.

# 6

## Working with Geospatial Information Data

### Exercise 1: Working with Remotely Sensed Data

In this lab, we will work with remotely sensed data, Landsat Thematic Mapper (TM) data, and use ERDAS-IMAGINE software to display image, image interpretation, and so forth. The following steps are needed for this lab.

To start ERDAS-IMAGINE type *imagine,* then open the viewer in ERDAS-IMAGINE. Within the viewer:

Click on File.

Click on Open.

Click on Raster.

Click filename to open and display the image or grid.

Display the Rocky Mountain National Park Landsat, Colorado, TM ETM+ (TM-7) image as a single band or band combination (use any band combination you like "band432 or b543, etc.") and observe the image carefully and report what you see. Look for vegetation, water, urban features, and so forth. Check layers information, map projection, and cell sizes. Is your image interpretation similar or different from other students? Can you distinguish among vegetation-forest type, water, trial, open area, meadow, and other land use and land cover types?

### Exercise 2: Derived Remote Sensing Data and Digital Elevation Model (DEM)

In this lab, we will work with derived remote sensing data such as vegetation indices, tassel cap transformation, and digital elevation model (DEM) data. You can use ARCINFO (GRID), ARCGIS, ARCVIEW, or ERDAS-IMAGINE software to display, interpret, and describe transfer remotely sensed data (Landsat TM) and DEM.

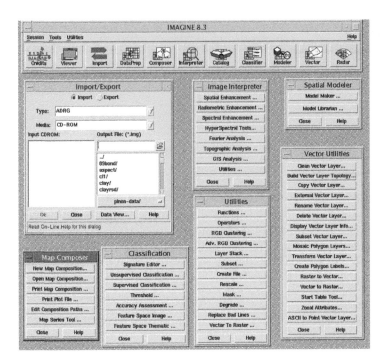

**FIGURE 6.1**
Layout for different functions for image processing and analysis using IMAGINE.

The DEM is a raster format data structure (grid cell, x and y). Raster data is a cellular-based data structure composed of rows and columns. You will work on the directory C:\myR\romodata. Then change the directory to romodata. Copy the DEM files to your own space directory that you created from the previous lab.

Then, use the Import command in ARCINFO to convert the export files to grid file format, that is:

```
arc: import grid elv.e00 elv
```

To start ERDAS-IMAGINE type *imagine* at the Unix prompt (%), or click on the imagine icon in PC and open the viewer in ERDAS-IMAGINE. Within the viewer:

Click on File.

Click on Open.

Click on Raster.

Click filename to open and display as a grid.

Display the grid as a single band and observe the grid carefully and report what you see. To examine the information on grid, click on Utility and Layer Info in your viewer window.

To display the grid in ARCINFO:

```
Arc: display 9999
Arc: gridgrid: mape elvgrid: gridpaint elv value linear
nowrap gray
```

Describe the grid file for information and statistics using the following command:

At Grid or ARC prompt:

Arc: describe elv

Read the information about the grid, data structure, statistical information (i.e., size the area covered by grid, geographic coordinates, pixel size). Also, read projection information (i.e., type of projection "UTM," and other information).

## Deriving Slope and Aspect from DEM Data

If we have a DEM called elv, we can derive slope and aspect in these steps:

```
Grid: slp = slope (elv, degree) "degree is optional, may use
percent"
Grid: asp = aspect (elv)
```

Describe the new grids, and report the statistics about slope and aspect you generate.

Sometimes when dealing with vegetation or ecological studies, we may need to transform aspect data using the absolute value from due south (180°; high solar radiation). The following example shows how to use absolute value:

```
Grid: absasp = abs(180 - elv),
```

Repeat your work on the different output between the two aspect files you generate in previous steps. Describe the new aspect and report the statistics.

## Resample GRID

In order to keep all your GIS layers in the same cell size, we can resample the raster data (remotely sensed data or DEM) to different cell sizes (pixel sizes). In this example, we resample DEM data from 30 m × 30 m to 20 m × 20 m.

```
Grid: elv20 = resample (elv, 20)
```

Describe the new grid and repeat your work on other DEM files. (You may select different cell sizes, i.e., 10, 15, 60, 90, 120).

For this lab, write a brief report to describe the differences of each grid file you used. What are the results of your interpretation, and what is your conclusion (what did you learn from this lab)?

---

### Exercise 3: Geospatial Information Data Extraction

Note: This lab is to show how to use the geospatial information data and extract information you need. It is not necessary that you will need to do this in this lab; however, this is just to help to understand the necessary steps needed for extraction of the data.

In this lab, we will extract topographic data (elevation, slope, and aspect), and remotely sensed data (TM) for the purpose of geospatial statistical modeling and mapping (i.e., develop trend surface model). Use ARCINFO/ARCGIS or ERDAS-IMAGINE software for this lab.

The following steps are needed: Copy the DEM and Landsat TM data files from the following path to your own "myR" directory that you created from the previous lab: /myR/romodata. Then, copy the following files from directory romodata:

| | |
|---|---|
| elv.e00 | elevation file in arc export format (grid- raster data) |
| romobond.e00 | boundary file in arc export format (cover- vector data) |
| romo.img | Landsat TM-7 image (raster data) |
| romo-data | Files contain x and y coordinates (ascii format) |
| xycord.aml | AML program to extract data based on x and y coordinate locations |

```
arc: import grid elv.e00 elv
arc: import cover romobond.e00 romobond
```

To start ERDAS-IMAGINE type *imagine* at the Unix prompt (%), then open the viewer in ERDAS-IMAGINE. Within the viewer:

Click on File.

Click on Open.

Click on Raster.

Click filename to open and display as an image or grid.

Display the image or grid, observe carefully, and report what you see. To examine the information on image or grid, click on Utility and Layer Info in your viewer window.

To display the grid in ARCINFO as:

```
Arc: display 9999
Arc: grid grid: mape elvgrid: gridpaint elv value linear
nowrap gray
grid: describe elv
```

### Deriving SLOPE and ASPECT from DEM Data (ELEVATION)

From elevation data, elv, derive slope and aspect as in these steps:

```
Grid: slp = slope (elv, degree)
Grid: asp = aspect (elv), then transform aspect to 180
```

Transform aspect data using the absolute value from due south (180°; high solar radiation). The following example shows how to use absolute value:

```
Grid: absasp = abs(180 - asp)
```

Check for data structure and cell sizes? If they are not the same cell sizes, then use resample in GRID as shown next.

### Resample GRID

In this example, we resample DEM data from 30 m × 30 m to 20 m × 20 m.

```
Grid: elv 25 (elv, 25)
```

After copying the all files, you can derive a new data set for the purpose of developing trend surface. In the following steps we can derive a vegetation index or tasseled cap (measure of greeness and yellowness [brightness], this function used for vegetation and soil studies) from Landsat TM, Multispectral Scanner (MSS), or SPOT images and use the data as a new layer in GIS. In ERDAS-IMAGINE:

Click on Image Interperter.

Click on Spectral Enhancemnet.

Click on Tasseled Cap, or if you are developing a vegetation index, then click on Indices. You will find different forms of vegetation indices (i.e., Veg. Index, NDVI, TNDVI, etc.) when using ERDAS-IMAGINE.

Be sure to put your output image in the same directory (on your disk space). Display the new images and report your results (statistics and other info). Compare the results with the original image.

In order to extract the spatial information from TM and radar data, elevation, slope, and aspect, we will convert the image data to grid file format. We can do that in ERDAS-IMAGINE using the Export function (you should know how to that from previous lab), or convert the image to grid in ARC:

```
Arc : imagegrid romo.img romo
```

The imagegrid procedure converts an image of Landsat TM-7 bands (room) to a grid file format readable in GRID-ARCINFO. The output has seven layers, or grids, named: romo_L1, romo _L2, romo _L3, romo _L4, romo_L5, romo_L6, and romo _L7.

### Select Area of Interest (Study Site)

After converting the Landsat TM imagery to a grid file format, one can produce a boundary file to represent a study site (area of interest). If the boundary file is in a vector format, then in ARCINFO use the latticeclip command as:

```
arc: latticeclip romo_L1 romobond romoarea_L1
```

The grid romo_L1 is clipped using the cover romobond and creates an output grid called romarea. Do this for all seven bands (grid format).

### Data Extraction

To develop a spatial model, we need to extract the data from the various GIS data layers associated with the field data. The following is an example of an Arc Macro Language (AML) program called xycord.aml used for data extraction. The program output is an ASCII text file containing the x, y coordinates and the grid values associated with the coordinates of the field data. In this example, we want to extract the digital number (DN) values for Landsat TM band-1 (romo_L1, grid file format) associated with the x, y coordinates stored in the ASCII text file called romodata.asc.

```
grid
display 9999
mape romo_L1                                    [grid file]

gridpaint romo_L1 value linear nowrap gray      [grid file]

&sv unit1 = [open romodata openstat1 -read]
[x, y-coordinates file to read in the program]
```

```
&sv unit2 = [open romo_L1.asc openstat2 -write] [output file to
write]
&sv record = [unquote [read %unit1% readstat]]
&sv plot = [extract 1 %record%]
&type plot# = %plot%
&sv x = [extract 2 %record%]
&type x-coordinate = %x%
&sv y = [extract 3 %record%]
&type y-coordinate = %y%
&sv value = [show cellvalue romo_L1 %x% %y%]    [grid file]
&type band value = %value%
&sv plotvalue = [quote %plot% %value%]
&sv writestat = [write %unit2% 'PLOT# Band1']   [Column label]
&sv writestat = [write %unit2% %plotvalue%]
&do count = 1 &to 207 &by 1
&sv record = [unquote [read %unit1% readstat]]
&sv plot = [extract 1 %record%]
&type plot# = %plot%
&sv x = [extract 2 %record%]
&type x-coordinate = %x%
&sv y = [extract 3 %record%]
&type y-coordinate = %y%
&sv value = [show cellvalue romo_L1 %x% %y%]    [grid file]
&type band value = %value%
&sv plotvalue = [quote %plot% %value%]
&sv writestat = [write %unit2% %plotvalue%]
&end
&sv closestat1 = [close %unit1%]
&sv closestat2 = [close %unit2%]
&term 9999 &popup romo_L1.asc                   [output file]
quit
```

To run this AML, use the command:

```
arc: &run xycord.aml
```

This AML could be used to extract the digital data associated with the field data from each of the GIS data layers, by changing the appropriate file names highlighted in bold.

For this lab, write a brief report to describe all the steps you worked on, the difference among each step, and the grid files you used. What are the results of your interpretation, and what is your conclusion (what did you learn from this lab)?

The following is an example on how the data output looks when using the xycord.aml program to extract the data for purpose of spatial modeling. The data are in Microsoft Excel format.

| Plot | X-UTM | Y-UTM | Elevation | Slope | ABS-aspect | TM Band-1 |
|------|-------|-------|-----------|-------|------------|-----------|
| 1 | 446862 | 4469209 | 2463 | 14.58181 | 31.9081 | 69 |
| 2 | 446774 | 4469317 | 2847 | 29.88589 | 16.8584 | 64 |
| 3 | 447147 | 4469658 | 2636 | 23.56723 | 46.5482 | 68 |
| 4 | 446511 | 4469750 | 2799 | 22.46655 | 130.9144 | 51 |
| 5 | 448319 | 4468602 | 2561 | 5.788832 | 99.46232 | 57 |
| 6 | 448641 | 4468469 | 2703 | 32.853 | 160.3959 | 51 |
| 7 | 448914 | 4468630 | 2690 | 14.9105 | 159.8637 | 46 |
| 8 | 449174 | 4468466 | 2627 | 9.736777 | 150.9454 | 55 |
| 9 | 447804 | 4468814 | 2561 | 5.788832 | 99.46232 | 57 |

### Steps for Converting the Geospatial Model to a Thematic Map Product

The spatial models developed in this class will consist of two model components, a trend surface model describing the large-scale spatial variability and a surface of residual describing the small-scale spatial variability. To convert the resulting spatial model to a map product, the following steps may be useful.

Sometimes when we are developing a trend surface model to describe the large-scale variability, the X-UTM and Y-UTM are important variables to use in the model we need to create a grid of X-UTM and Y-UTM coordinates. The following grid command uses the DEM elv and creates grids called X-UTM and Y-UTM.

```
Grid: setcell elv
Grid: setwindow elv
Grid: UTM-X = $$XMAP
Running... 100%
Grid: UTM-Y = $$YMAP
Running... 100%
```

It is recommended to use the describe command in ARC or GRID. This provides useful information about a surface (grid).

```
Grid: describe UTM-X

            Description of Grid UTM-X
Cell Size = 25.000      Data Type: Floating Point
Number of Rows = 1367
Number of Columns = 2091

BOUNDARY                STATISTICS
Xmin = 573262.069       Minimum Value = 573274.562
Xmax = 625537.069       Maximum Value = 625524.562
Ymin = 4132888.119      Mean = 599399.562
Ymax = 4167063.119      Standard Deviation = 15090.491
```

We need to clip the residual (a surface called "risd.asc" resulted from S-Plus programming by converting to a grid using the asciigrid command in ARC):

```
Arc: asciigrid risd.asc risd float
```

Then, use the surface risd to clip with vegetation map rmnp8000-veg and produce an output rmnp8000-risd to be used as input with the final trend surface model:

```
Arc: latticeclip risd romobond romo-risd
```

We can use addition or subtraction or any mathematical operation in GRID. The following example generates a grid estimating the number of native species using a trend surface model to describe the large-scale spatial variability and a surface of residuals (rmnp8000-risd) to describe the small-scale spatial variability.

```
Grid: Native = -454.2 + 0.34820223 * rmnp8000_L1 - 0.39359571
* rmnp8000_L3
+ 0.10718073 *rmnp8000_L5 + 0.14523950 * rmnp8000_L6 -
0.168334567 * rmnp8000_L7 + 0.00585931 * dem8000 - 0.00024156
* UTM-X + 0.00011576 * UTM-Y + rmnp800 0-risd
Grid: describe Native

 Description of Grid NATIVE

Cell Size = 30.000  Data Type: Floating Point
Number of Rows = 652
Number of Columns = 933


BOUNDARY                STATISTICS
Xmin = 425760.000       Minimum Value = -26.209
Xmax = 453750.000       Maximum Value = 30.828
Ymin = 4459810.000      Mean = 1.389
Ymax = 4479370.000      Standard Deviation = 5.016

COORDINATE SYSTEM DESCRIPTION
Projection   UTM
Zone         NOT DEFINED
Datum        NONE
Units        METERS      Spheroid       CLARKE1866
Parameters:
longitude -105 0 0.00 latitude 40 0 0.000
```

Sometimes when we are developing these models, the final surface may have values that exceed some specified upper and lower bound (i.e., 0 to 1).

We can correct for these problems using the following steps in ARCINFO. In the following example, all negative values in the grid "native" are set to 0.

```
Grid: docell
:: if( native < 0 )
:: tmp = 0
:: else native1 = native
:: end
```

The next step is to delete the grid "tmp" using the following command in ARCINFO using grid:

```
Grid: docell
:: if( pino-sand2 < 0 )
:: tmp = 0
:: else pinon-sand3 = pinon-sand2
:: end
```

Or use the "con" statement in grid as follow:

```
Gird: native1 = con (native < 0, 0, native)
```

If you are developing a surface based on probability (present and absence, i.e., predict weed,) then we can use "con" as follows:

```
Weedp = con (weed < 0, 0, con (weed > 1, 1, weed)))
```

Next step is to delete the grid "tmp" using the following command in ARCINFO, then display grid "pinon-sand3" in ARCVIEW.

```
Arc: kill tmp all
```

## Working with Vegetation Indices and Tasseled Cap Transformation

### *Vegetation Indices*

Measurements of the amount and condition of vegetation are based on an analysis of remote sensing spectral measurements using vegetation indexes. The objective is to reduce the multiple bands of data to a single number per pixel that will predict or assess such vegetation characteristics as canopy opening, percent cover, biomass, productivity (phytomass), leaf area index (LAI), or amount of photosynthetically active radiation (PAR) consumed (Larsson 1983). There are many vegetation indices that can be used for the purpose of vegetation analysis studies. For example, if we are using Landsat

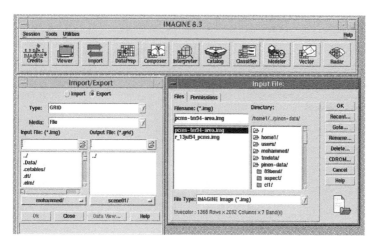

**FIGURE 6.2**
Using Import/Export in ERDAS-IMAGINE.

TM Data, MODIS, ASTER, SPOT HRV, or Advanced Very High Resolution Radiometer (AVHRR), one can use one of the following simple forms to calculate the vegetation index:

$$NDVI_{TM} = [(TM4 - TM3)/(TM4 + TM3)] \qquad (6.1)$$

$$NDVI_{HRV} = [(XS3 - XS2)/(XS3 + XS2)] \qquad (6.2)$$

$$NDVI_{AVHRR} = [(IR - red)/(IR + red)] \qquad (6.3)$$

### *Tasseled Cap*

The most important vegetation index is the tasseled cap transformation developed by Kauth and Thomas (1976). This index is based on Gram–Schmidt sequential orthogonalization techniques that produce an orthogonal transformation of the original four bands of multispectral scanner (MSS) data space to a new four-dimensional space (Jensen 1996). The output from tasseled cap data has a shape like a cap. It has been used most commonly in agricultural research (vegetation vs. soil analysis). Kauth and Thomas reported that the new transformation identifies four new axes including the soil brightness (SBI), the green vegetation index (GVI), the yellow stuff index (YVI), and a nonsuch index (NSI) associated with atmospheric effects. In general, the first two indexes contain most of the scene information (95% to 98%). Kauth and Thomas found that nearly all (98%) of the variance in bare soil spectra from several different soil types could be explained by the

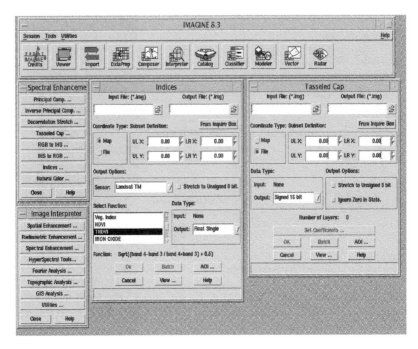

**FIGURE 6.3**
Using the Vegetation Index and Tasseled Cap Function in ERDAS-IMAGINE.

soil brightness index. The study concluded that bare soils would lie in a line parallel to the brightness axis and that a globally valid soil line exists that could be applied to Landsat MSS agricultural scenes. However, greenness is an orthogonal deviation from the mean soil line and is used as a measure of the green vegetation present. Jensen (1996) pointed out that the further the distance perpendicular to the soil line, the greater the amount of vegetation present within the field of view of a pixel.

Researchers in remote sensing were able to evaluate the Landsat 4 and 5 Thematic Mapper data to determine if tasseled cap provides as useful vegetation index as the MSS data. Crist and Cicone (1984) found that "the coefficient of the first two components are, for the TM bands which sample the same spectral regions as the MSS bands (TM bands 2, 3, and 4), comparable to those which define *greenness* and *brightness*." They found that the six-band TM data are dispersed into a three-dimensional space and, more precisely, define the two (Jensen 1996). Fully vegetated areas define the plane of vegetation, whereas bare soils data fall in a plane of soils. Between the two are data with partially vegetated plots where both vegetation and soil are visible. Thus, during a growing season an agricultural field is expected to begin on the plane of soils, migrate through the transition zone, arrive at the plane of vegetation near the end of crop of soils during harvest or senescence. Also, the study of Crist and Cicone (1984) identified a third component that is related to soil feature: moisture status. Thus, an important new source of soil information is available through the inclusion of the middle-infrared bands of the TM. The study further identified the fourth tasseled cap parameter as being haze. The information derived from this parameter may be used to dehaze Landsat TM data (Lavreau, 1991).

### Develop Thematic Layer in ARCVIEW or ARCMAP

#### *VIEWS (Working Only with ARCVIEW)*

Within a project, a view is a picture (Figure 6.4) of geographic data (i.e., coverage such as a soils map or a map of vegetation). This view can be displayed, queried, analyzed, explored, and manipulated. The view is linked to the data, but the data (coverage) are not changed when you manipulate a view. If the data changed, the associated view will automatically reflect those changes. A view is actually a collection of one or more themes. A theme is analogous to the information in a single coverage. A vegetation coverage has information about vegetation. A soil texture coverage has information on soil texture. These are examples of themes. Several themes can be put into one view. These steps are similar to plotting different coverage on top of one another (in different colors) when using ARCPLOT in ARCINFO. The difference is that views are saved (as part of a project) for later use.

To add a theme to a view, you need to do the following steps (also see Figure 6.4):

1. Click on Views, then the New button. A window named View 1 will be created.
2. Select View. Then select. Add theme. Change your directory or disk drive, select a coverage (or theme). Then click the title gray box on the legend of the view to activate the theme.
3. Go to View, then Properties. Change the name of the view to something meaningful. Also change map unite to meters or feet (do not change anything else here).
4. Double-click on the colored legend box; the Legend Editor will appear. Choose a Field (from the theme's attribute table) that you wish to classify on (probably the "code" item).
5. Next select Classify and pick one of the methods (you may end up using your own values, or specify how many classes you want), then put your own values in (if necessary).
6. To edit the color or fill pattern, double-click on one of the specific legend patterns and the palettes window will appear. Next click on

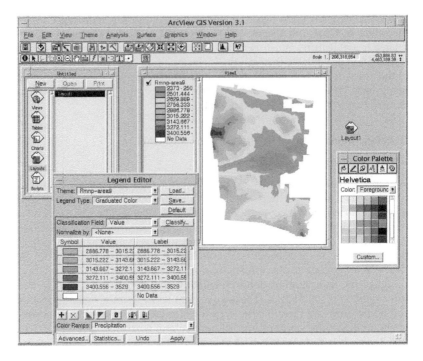

**FIGURE 6.4**
An example of displaying the viewer and other information needed to add a theme.

the Labels circle and change these numbers to whatever you want to appear in the legend area (i.e., urban, forest, road, range of numbers, etc.).

Finally, hit the Apply button to activate all legend edits. You can save your new legend as a pinon-sand.avl file by hitting the Save button. (Note: You can load previously saved [i.e., pinon-sand.avl or pinon-elv.avl] files to use with other views.) Finally, close the Legend Editor Window when you done.

7. Add attribute information (code values) to the picture by selecting Theme, then Properties, then Text Labels, then Label Field (pick your code item here). Hit OK, then Theme, and then Auto label.

8. To edit the text (fonts, sizes) just created, select Edit, then Select all graphics, then Window, then Show Symbol Palette. Then select the ABC button on the palette.

9. To save the view, save project.

10. To print the view, click File, then Print.

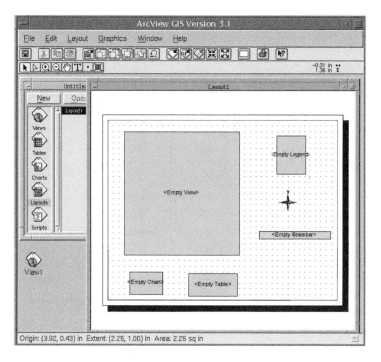

**FIGURE 6.5**

An example of a map layout with different frames, view, legend, scale bar, north arrow, and so forth.

We can also use the main menu icons: Identify, Hammer (query builder), Table Opener, Select, Legend Editor, Theme Properties, Add Theme, the Hand, Zoom In, Zoom Out, Text, and others. You need to experiment with them to better learn their functions.

## Map Layout

A layout is a picture that can be created of the views, with scale bars, north arrows, legend, title, and others (see Figure 6.5). This layout can then be saved and sent to a printer. To create a Layout, do the following:

1. Click on Layouts, then the New button. A layout named Layout 1 will be created.
2. Add frames to the layout. This icon(s) are on the second row, far right: view, scale bar, legend, north arrow. To edit them, select and double-click them. We can resize the frames (this is one way to adjust the your scale).

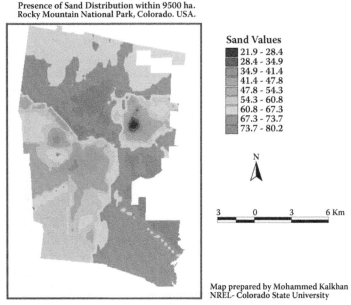

**Presence of Sand Distribution within 9500 ha.
Rocky Mountain National Park, Colorado. USA.**

**Sand Values**
21.9 - 28.4
28.4 - 34.9
34.9 - 41.4
41.4 - 47.8
47.8 - 54.3
54.3 - 60.8
60.8 - 67.3
67.3 - 73.7
73.7 - 80.2

N

3          0          3          6 Km

Map prepared by Mohammed Kalkhan
NREL- Colorado State University

**FIGURE 6.6**
Example on developing a thematic map using layout in ARCVIEW.

3. Add text to the layout with the T icon.

4. Choose layout Properties to give a name to layout.

5. Save the layout.

6. To print the layout, click File , then Print.

Using this feature, one can create templates for other views. (See Figure 6.5 and Figure 6.6).

## References

Crist, E. P., and R. C. Cicone. 1984. Application of the tasseled cap concept to simulated thematic mapper data. *Photogrammetric Engineering & Remote Sensing* 50:343–352.

Jensen, J. R. 1996. *Introductory Digital Image Processing: A Remote Sensing Perspective*, 2nd ed. Upper Saddle River, NJ: Prentice Hall.

Kauth R. J., and G. S. Thomas. 1976. The tasseled cap—A graphic description of the spectral-temporal development of agricultural crops as seen by Landsat. In *Proceedings of the Symposium on Machine Processing of Remotely Sensed Data*, 41–51. West Lafayette, IN: Laboratory for Application of Remote Sensing.

Larson, D. W. 1983. Morphological variation and development in Ramalina menziesii. *American Journal of Botany* 79, 668–681.

Lavreau, J. 1991. De-hazing Landsat Thematic Mapper images. *Photogrammetric Engineering & Remote Sensing* 57(10):1297–1302.

# *Index*

**A**

ARCINFO software, 29, 31–32, 148–151
Arc Macro Language, 150, 151
ASTER (Advanced Spaceborne Thermal Emission and Reflection radiometer), 13, 14

**B**

Bi-Moran's *I* statistic, 73–75, 133–134
Binary classification trees (BCTs)
  overview, 97
  vegetation map classes, for predicting, 97–100

**C**

Classification and regression tree (CART), 107
Cokriging, 100, 102–104. *See also* Kriging
Cross-correlation statistics, 67
Cross-validation model, 91

**D**

Data collection. *See also* Sampling methods
  error sources, potential, 39
  intervals, 40
  methods, 39
  nominals, 40
  ordinals, 40
  ratios, 39, 40
Digital elevation model (DEM), 1, 145–148

**E**

Earth Observing System (EOS) satellites, 13. *See also specific satellites*
Envelope model, 108

ERDAS-IMAGINE software, 27–29, 145, 150
European Remote Sensing Satellite (ERS-1), 15

**G**

GARP model (genetic algorithm for rule-set prediction model), 105–106
Geary's *C* statistic, 66, 67, 71–72, 84, 132–133
Geographic information systems (GIS)
  advantages of each type, 26
  raster data type, 1, 25, 26
  real-world representations, 25
  vector type, 25–26
Geospatial models for presence and absence data
  genetic algorithm for rule-set prediction model, 105–106
  logistic regression models, 106–107
  Maxent model, 106
  overview, 104–105
Global change forecasting, 9
Global positioning systems (GPS)
  applications, 34–35
  augmentation systems, 33
  civilian use, 35
  control and monitoring stations, 32
  maintenance of, 33
  overview, 32
  receivers, 32
  satellites, 32, 33, 34
  tagging coordinates, 35
  services, 33
  usage, 32–33
Grand Staircase-Escalante National Monument (GSENM)
  case study, 64, 65

**I**

IKONOS satellite
    characteristics, 5
    overview, 4–5
    PNP, use with, 48
    specifications, 5
    weight of, 5
Inverse distance weighting, 67–69

**K**

Kauth, R. J., 22, 23, 24
Kriging, 84
    assumptions, 85
    case example, 87–90
    cokriging, 100, 102–104
    ordinary, 85
    simple, 85, 86–87
    universal, 85–86, 87

**L**

Landsat Multispectral Scanner
        (MSS), 1, 22
    bandwidths, 3–4
    instantaneous field of view, 3
    PNP, use with, 48
    system characteristics, 3
Landsat Thematic mapper
    accuracy, 45
    imagery, conversion of, 27, 28,
        29, 31
Lasers, 18, 19. *See also Lidar*
Lidar
    DIAL, 17
    Doppler lidars, 17–18
    lasers, use of, 18
    principles of use, 17
    range finders, 17
    systems, difference between, 18
    usage, description of, 18–19
Linear correlation statistics, 63–64, 65
Logistic regression models, 106–107

**M**

Maps, layout of, 159–160
Maxent model, 106

**MODIS** (Moderate Resolution Imaging
        Spectroradiometer), 9, 10, 12, 13
Moran's $I$ statistic, 65–66, 67, 69–71,
        131–132, 137

**N**

NAVSTAR, 34

**O**

ORBIMAGE (GeoEye)
    overview, 5
    specifications, OrbView-2, 6
    specifications, OrbView-3, 6
Ordinary least squares
    applications, 81
    case example, 93–94
    methodology, 81–82, 83
    overview, 81

**P**

Pearson correlation coefficient, 64
Pixel nested plot (PNP)
    Case study with R, 129
    biodiversity measurement
        with, 47–49
    design of, 49–50
    overview, 46–47
Plot designs
    circular plots, 49–50
    dimensions, recording, 51
    overview, 49
    rectangular plots, 50–51
    shape, determining, 49
    size, 51–52
    vegetation studies, use with, 51–52

**Q**

QuickBird satellites 6–7, 48

**R**

R statistical package
    advantages of, 116, 117
    case study, 128–134
    data comparison with, 122–123

defining R, 116
development, 116
environment, 117
graphics with, 119, 121–122
histograms with, 122
linear correlations, testing
  for, 126–127
logical vectors with, 121
overview, 116
plotting with, 126
programming with, 120
regression analysis with, 127–128
SAR models, 139–140
spatial autocorrelation, 136–137
statistics with, 119, 124
start-up, 118–119
suite, integrated, 117
trend surface analysis, 134–136
utilizing, 118
variables, 124–125
variogram model, 140–143
Radar
  history of use, 15
  overview, 15–16
RADARSAT, 48
raster format data structures, 27
Remotely sensed data. *See also*
    *specific satellites and data*
  history of, 2t
  overview, 1
Resolution, spatial, 1

**S**

Sampling methods
  accuracy, 45
  cluster sampling, 44
  multiphase, 41, 44–45
  nonaligned systematic
    sampling, 44
  nonrandom sampling, 41
  overview, 40
  pixel nested plot (PNP). *See* pixel
    nested plot (PNP)
  probability sampling, 40–41
  simple random, 41–42
  spatial sampling. *See* Spatial
    sampling
  stratified, 42–43

Scale. *See also* Spatial sampling
  geostatistical applications, 57–58
  large, 57
  medium, 57
  overview, 57
  small, 57
Seasat oceanographic satellite, 1
Semivariogram, 84, 86. *See also* Kriging
Spatial autoregressive (SAR) models
  case example, 94–96,
    136–137, 138–139
  overview, 91–92
  sampling units, 92
  steps in, 91–92
Spatial correlation statistics
  applications, 65
  bi-Moran's $I$ statistic, 73–75, 133–134
  cross-correlation statistics, 67
  Geary's $C$ statistic, 66, 67, 71–72,
    84, 132–133
  inverse distance weighting, 67–69
  Moran's $I$ statistic, 65–66, 67,
    69–71, 131–132, 137
  overview, 65
Spatial modeling, 43
Spatial patterns
  analysis of, 61–63
  considerations regarding, 59
  overview, 59
  point patterns, 59–63
  quadrant count method, 63
  variance to mean ratio, 63
Spatial sampling
  errors, potential, 58
  overview, 58
  predictions based on, 58–59
  variability in, 58–59
SPOT (System Probatori D'Observation
    de la Terre) satellites, 7–8, 48
Stepwise regression
  case example, 80–81
  methodology, 79–80
  overview, 79–80

**T**

Tassled cap transformation,
    22–24, 155–157
Thematic data layers, 26, 157–59

Thematic Mapper (TM5), 1, 22
Thomas, G. S., 22, 23, 24

**V**

Variogram model, 83–85, 87–90, 140–143.
    *See also* Kriging
Vegetation indices
  accuracy, 21
  advantages of, 20–21
  analysis studies, 154–155
  calculations/construction of, 20
  disadvantages of, 20
  measurements, 154
  newer generations of, 2
  overview, 19–20
  usages, 21
visible near-infrared (VNIR) telescopes, 13

Printed and bound by CPI Group (UK) Ltd, Croydon, CR0 4YY

01/11/2024

01782618-0009